I0034337

# Advanced Techniques of Analytical Chemistry

# (Volume 1)

Edited by

**Anju Goyal**
*Department of Pharmaceutical Chemistry,*
*Chitkara College of Pharmacy,*
*Chitkara University*
*Punjab*
*India*

&

**Harish Kumar**
*Govt. College of Pharmacy,*
*Rohru, Shimla,*
*Himachal Pradesh*
*India*

# Advanced Techniques of Analytical Chemistry

*Volume # 1*

Editors: Anju Goyal and Harish Kumar

ISBN (Online): 978-981-5050-23-3

ISBN (Print): 978-981-5050-24-0

ISBN (Paperback): 978-981-5050-25-7

© 2022, Bentham Books imprint.

Published by Bentham Science Publishers Pte. Ltd. Singapore. All Rights Reserved.

# BENTHAM SCIENCE PUBLISHERS LTD.
## End User License Agreement (for non-institutional, personal use)

This is an agreement between you and Bentham Science Publishers Ltd. Please read this License Agreement carefully before using the ebook/echapter/ejournal (**"Work"**). Your use of the Work constitutes your agreement to the terms and conditions set forth in this License Agreement. If you do not agree to these terms and conditions then you should not use the Work.

Bentham Science Publishers agrees to grant you a non-exclusive, non-transferable limited license to use the Work subject to and in accordance with the following terms and conditions. This License Agreement is for non-library, personal use only. For a library / institutional / multi user license in respect of the Work, please contact: permission@benthamscience.net.

## Usage Rules:

1. All rights reserved: The Work is the subject of copyright and Bentham Science Publishers either owns the Work (and the copyright in it) or is licensed to distribute the Work. You shall not copy, reproduce, modify, remove, delete, augment, add to, publish, transmit, sell, resell, create derivative works from, or in any way exploit the Work or make the Work available for others to do any of the same, in any form or by any means, in whole or in part, in each case without the prior written permission of Bentham Science Publishers, unless stated otherwise in this License Agreement.
2. You may download a copy of the Work on one occasion to one personal computer (including tablet, laptop, desktop, or other such devices). You may make one back-up copy of the Work to avoid losing it.
3. The unauthorised use or distribution of copyrighted or other proprietary content is illegal and could subject you to liability for substantial money damages. You will be liable for any damage resulting from your misuse of the Work or any violation of this License Agreement, including any infringement by you of copyrights or proprietary rights.

## *Disclaimer:*

Bentham Science Publishers does not guarantee that the information in the Work is error-free, or warrant that it will meet your requirements or that access to the Work will be uninterrupted or error-free. The Work is provided "as is" without warranty of any kind, either express or implied or statutory, including, without limitation, implied warranties of merchantability and fitness for a particular purpose. The entire risk as to the results and performance of the Work is assumed by you. No responsibility is assumed by Bentham Science Publishers, its staff, editors and/or authors for any injury and/or damage to persons or property as a matter of products liability, negligence or otherwise, or from any use or operation of any methods, products instruction, advertisements or ideas contained in the Work.

## *Limitation of Liability:*

In no event will Bentham Science Publishers, its staff, editors and/or authors, be liable for any damages, including, without limitation, special, incidental and/or consequential damages and/or damages for lost data and/or profits arising out of (whether directly or indirectly) the use or inability to use the Work. The entire liability of Bentham Science Publishers shall be limited to the amount actually paid by you for the Work.

## General:

1. Any dispute or claim arising out of or in connection with this License Agreement or the Work (including non-contractual disputes or claims) will be governed by and construed in accordance with the laws of Singapore. Each party agrees that the courts of the state of Singapore shall have exclusive jurisdiction to settle any dispute or claim arising out of or in connection with this License Agreement or the Work (including non-contractual disputes or claims).
2. Your rights under this License Agreement will automatically terminate without notice and without the

need for a court order if at any point you breach any terms of this License Agreement. In no event will any delay or failure by Bentham Science Publishers in enforcing your compliance with this License Agreement constitute a waiver of any of its rights.

3. You acknowledge that you have read this License Agreement, and agree to be bound by its terms and conditions. To the extent that any other terms and conditions presented on any website of Bentham Science Publishers conflict with, or are inconsistent with, the terms and conditions set out in this License Agreement, you acknowledge that the terms and conditions set out in this License Agreement shall prevail.

**Bentham Science Publishers Pte. Ltd.**
80 Robinson Road #02-00
Singapore 068898
Singapore
Email: subscriptions@benthamscience.net

**BENTHAM SCIENCE**

# CONTENTS

# FOREWORD

The classical analytical techniques include gravimetric, volumetric, and titrimetric methods; on the other hand, instrumental techniques involve ultraviolet-visible (UV-Vis), infrared (IR), and near-infrared (NIR) spectrophotometry fluorimetry, atomic spectroscopy (absorption/emission), electroanalytical chromatography.

The current volume of a book titled "**Advance Techniques of Analytical Chemistry-Volume-1**" is about techniques depending on the nature of reactions such as Acid-Base titrations which involve the reaction of an acid and a base. Redox titrations include redox reaction between analyte and titrant as the key reaction, Non-aqueous titrations, Complexometric titration which involves the formation of a colored complex compound, and some miscellaneous methods like Diazotisation Titrations, Kjeldahl Method and Oxygen flask combustion method, *etc.*

**Advance Techniques of Analytical Chemistry-Volume-1** covers a number of important analytical chemistry topics. It does not, however, cover spectroscopic such as mass spectroscopy, IR or NMR as well as chromatographical techniques such as TLC, HPLC, HPTLC and other advanced techniques because these qualitative applications will be covered adequately incoming volumes *i.e.* Volume-II and Volume-III.

We are happy to recommend this book to students and researchers as there is more material than anyone can cover in one semester; I hope that the diversity of topics will meet the needs of different instructors.

**Prabhakar Kumar Verma**
M.Pharm, Ph.D. (UDPS Nagpur)
Associate Professor of Pharmaceutical Chemistry
Department of Pharmaceutical Sciences,
M.D. University,
Rohtak-124001
Haryana
India

# PREFACE

The main purpose of this book is to cover the basic concept of volumetric titrations. Modern Advance Techniques of Analytical Chemistry essentially involves, as a necessary integral component, even greater horizons than the actual prevalent critical analysis of not only the active pharmaceutical substances but also the secondary pharmaceutical product(s) *i.e.*, the dosage forms having the either single or multi-component formulated product. The fundamental reasons for this sudden legitimate surge in the newer evolving methodologies in the 'analysis of drug substances' are perhaps due to the tremendous growth in the progress of 'medicinal chemistry' towards achieving one ultimate objective which is to obtain 'better drugs for a better world'. With the advent of computer-aided-drug modeling (CADM) the critical, scientific, and faster approach to newer drug entities based on the biologically active prototypes, combinatorial chemistry, chiral chemistry, and biotechnology has paved the way towards more specific, potent, and above all, less toxic 'drugs' to improve the ultimate quality of life in humans.

Keeping in view the above astronomical growth in the design of complicated, specific, and highly active drug molecules, equally viable, rigorous, accurate, and precise analytical methods have been evolved with the time which has now occupied pivotal and vital positions in most of the Official Compendia *viz*., USP, BP, Int.P., Eur. P, IP, *etc.*, for the analysis of such compounds both in pure and dosage forms.

The present book on 'Advance Techniques of Analytical Chemistry' caters for the much-needed handbook and reference book, which is current about the esteemed philosophy of analytical chemistry, obvious solid support towards drug discovery, development, stability studies, bioavailability, and pharmacokinetic studies, and above all the quality assurance of pure drugs together with their respective dosage forms.

The textbook on 'Advance Techniques of Analytical Chemistry' would enormously serve the undergraduates, postgraduates, researchers, analytical chemists working in the Quality Assurance Laboratories, new drug development, production and control, teaching, or regulatory authorities.

**Anju Goyal**
Chitkara College of Pharmacy
Chitkara University
Rajpura, Punjab
India

&

**Harish Kumar**
Govt. College of Pharmacy
Rohru, Shimla, Himachal Pradesh
India

# List of Contributors

| | |
|---|---|
| **Astha Sharma** | Laureate Institute of Pharmacy, Jawalamukhi, Himachal Pradesh, India |
| **Anju Goyal** | Chitkara College of Pharmacy, Chitkara University, Punjab, India |
| **Harish Verma** | Govt College of Pharmacy, Rohru, Shimla, Himachal Pradesh, India |
| **Komalpreet Kaur** | G.H.G Khalsa College of Pharmacy, Gurusar Sadhar, Ludhiana, India |
| **Madhukar Garg** | Chitkara College of Pharmacy, Chitkara University, Punjab, India |
| **Monika Gupta** | A.S.B.A.S.J.S. Memorial College of Pharmacy, Bela Punjab, India |
| **Navdeep Singh** | Laureate Institute of Pharmacy, Jawalamukhi, Himachal Pradesh, India |
| **Nidhi Garg** | Chitkara College of Pharmacy, Chitkara University, Punjab, India |
| **Payal Das** | Chitkara College of Pharmacy, Chitkara University, Punjab, India |
| **Rajendra Awasthi** | Amity Institute of Pharmacy, Amity University Uttar Pradesh, Noida, Uttar Pradesh, India |
| **Rajwinder Kaur** | Chitkara College of Pharmacy, Chitkara University, Punjab, India |
| **Ramninder Kaur** | Chitkara College of Pharmacy, Chitkara University, Punjab, India |
| **Sandeep Arora** | Chitkara College of Pharmacy, Chitkara University, Punjab, India |
| **Sapna Kumari** | Chitkara College of Pharmacy, Chitkara University, Punjab, India |
| **Shammy Jindal** | Laureate Institute of Pharmacy, Jawalamukhi, Himachal Pradesh, India |
| **Shivani Sharma** | RKSD College of Pharmacy, Kaithal-136027, Haryana, India |
| **Sunaina Aggarwal** | RKSD College of Pharmacy, Kaithal-136027, Haryana, India |
| **Upasna Kaushik** | RKSD College of Pharmacy, Kaithal-136027, Haryana, India |

# CHAPTER 1

# Introduction to Analytical Chemistry

**Sapna Kumari[1,*], Anju Goyal[1], Madhukar Garg[1] and Harish Verma[2]**

[1] *Chitkara College of Pharmacy, Chitkara University, Punjab, India*

[2] *Govt College of Pharmacy, Rohru, Shimla, Himachal Pradesh, India*

**Abstract:** Analytical chemistry, a branch of chemistry, deals with the analysis of substances. Identification of the constituents from the mixture or substance is called qualitative analysis while Quantitative Analysis deals with determining the purity of the constituents present in the mixture. This is also known as assay method. Various methods like volumetric analysis, gravimetric analysis, polarimetry, refractometry, photometry, fluorimetry, electrochemical methods, chromatographic methods and biological methods comes under the category of quantitative analysis. Volumetric titrations are the elementary procedures applied in the life sciences, pharmaceutics, industrial analysis, water pollution and clinical chemistry and are used to determine the amount of analyte with the standard solution. Main advantages of these methods are cost-effective, rapid and simple, while disadvantages include pH, temperature, and humidity sensitivity, indicator requirement, occurrence of human error, *etc.*

**Keywords:** Analytical chemistry, Endpoint, Indicator, Quantitative analysis, Volumetric titration.

## INTRODUCTION

Analytical chemistry is the branch of chemistry which deals with the analysis of substances. It is mainly divided into two categories:

(1) Qualitative Analysis

(2) Quantitative Analysis

Qualitative analysis is primarily concerned with the identification of the constituents present in a chemical substance or a mixture of substances.

Quantitative analysis is also primarily concerned with the exact determination of the purity of the number of constituents present in a chemical substance or a mixture of substances.

* **Corresponding Author Sapna Kumari**: Chitkara College of Pharmacy, Chitkara University, Punjab, India; Email: ms.sapnakumari92@gmail.com

**Anju Goyal & Harish Kumar (Eds.)**
**All rights reserved-© 2022 Bentham Science Publishers**

Quantitative analysis is carried out mainly for determining the purity of chemical substances. The method used for the determination of purity is called the assay method. There are many methods of quantitative analysis such as volumetric analysis, gravimetric analysis, polarimetry, refractometry, photometry, fluorimetry, electrochemical methods, chromatographic methods, and biological methods.

Volumetric analysis is also known as titrimetric analysis which is carried out in a routine manner and hence is discussed here [1].

## INTRODUCTION TO VOLUMETRIC TITRATIONS

Volumetric titrations are the basic techniques in chemistry that are applied in the life sciences, pharmaceutics, industrial analysis, water pollution, and clinical chemistry. Volumetric titrations or titrimetric analysis are the quantitative analytical procedures that are used to determine the amount of analyte with the standard solution. In this titration, the standard solution has been added gradually to the sample containing an unknown concentration of reactant until reactant is consumed (stochiometric completion). This is called equivalence point [2]. Indicators have been used to identify the endpoint. Popularly, these titrations are used in acid-base reactions.

The first method of Volumetric Analysis was devised and found by the French chemist Jean-Baptiste-Andre-Dumas as he was trying to determine the proportion of nitrogen combined with other elements in organic compounds. To ensure the conversion of the nitrogen compound into pure gas, the nitrogen compound was burnt in a furnace and passed along a furnace in a stream of carbon dioxide that is passed into a strong alkali solution. The mass of the nitrogen is calculated and occupied under known conditions of pressure and volume from the sample.

### Advantages And Disadvantages Of Volumetric Analysis

Titration is a method used to determine the concentration of an unknown sample solution by a recognized concentration of solute. The titrimetric method includes three types: volumetric, gravimetric, and coulometric titrimetric. The volumetric titration was used to determine the volume of solution with the known concentration involving a quantitative reaction with a substance solution to be analysed.

The Advantages of Volumetric Titration are as Follows:

• The major advantage of volumetric titration is that it is a simple and cost-

effective method.
- Volumetric titration is not sophisticated, so skill is not required to handle it.
- The volumetric titration is rapid and gives accurate results.
- It is a simple method compared to other types of titrations.
- The reaction can identify visually at equilibrium or endpoint.

The Disadvantages of Volumetric Titration are as Follows:

- Certain factors such as pH, temperature, and humidity may affect the titration results, as this is an open system.
- Human error may occur during processing and may affect accuracy.
- It requires an indicator for the reactions to occur.
- The major disadvantage of volumetric titration is that this can produce large volumes of chemical waste and that needs to be disposed of it.
- It requires reactions occurring in a liquid phase.

## WHAT IS TITRATION?

This is a basic laboratory method of quantitative chemical analysis used to determine the unknown concentration of a known reactant. In this method, volume measurements play a key role in titration, also known as volumetric analysis (Fig. **1**). A reagent called the titrant or titrator of a known concentration (a standard solution) and volume is used to react with a solution of the analyte or titrant whose concentration is not known [3, 4]. Using a calibrated burette or chemistry pipetting syringe to add the titrant.

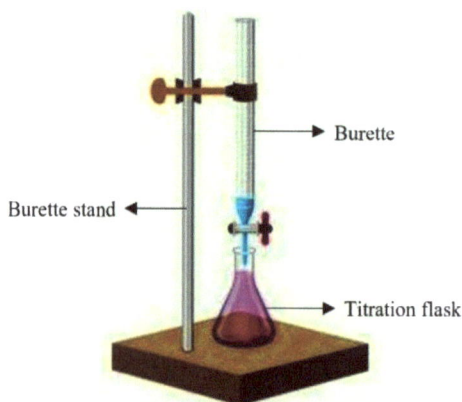

**Fig. (1).** Representing assembly of procedure of titration.

## Simple Titrations

Simple titration aims to find the concentration of an unknown solution with the help of the known concentration of another solution (Fig. **2**).

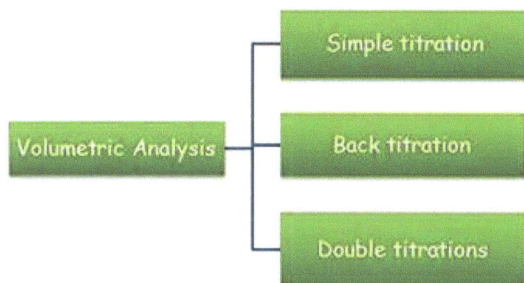

**Fig. (2).** Illustrating volumetric titrations.

## The Volumetric Analysis is Divided into the Following Types:

Let us take a solution of a substance 'A' of unknown concentration. We are provided with a solution of another substance 'B' whose concentration is known ($N_1$). We take a certain known volume ($V_2$ liters) of 'A' in a flask and start adding 'B' from the burette to 'A' slowly till all the 'A' is consumed by 'B'. This can be known with the aid of a suitable indicator, which shows colour change after the complete consumption of 'A'. Let the volume of B consumed be $V_1$ liter According to the law of equivalents, the number of equivalents of 'A' would be equal to the number of equivalents of 'B'.

Hence

$$N_1V_1 = N_2V_2$$

where $N_2$ is the concentration of 'A'.

Thus, using this equation, the value of $N_2$ can be calculated.

There are four types of simple titration, namely.

- Acid-base titrations
- Redox titrations
- Precipitation titrations and
- Complexometric titrations

## Types of Volumetric Titrations

There are various sorts of titrations whose goals are different from the others. The most common types of titrations in qualitative work are acid-base titrations, redox titrations, complexometric titration and precipitation titration (Fig. **3**).

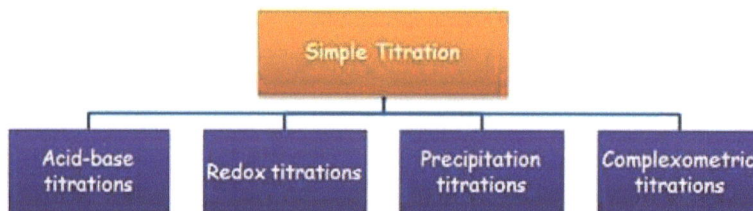

**Fig. (3).** Representing different types of simple titrations.

### *Acid-Base Titrations*

The main base of these types of titrations is a neutralization reaction that befalls between an acid and a base when mixed together into a solution (Fig. **4**). The acid is added to a burette and the burette was rinsed prior with the same acid, which was supposed to add to prevent contamination and dilution of acid being added in the same. volumetric flask, which should be rinsed properly with the same solution added in the same to prevent contamination or dilution of the base/alkali being measured. The solution in the volumetric flask is often a standard solution having a known concentration.

However, the solution whose concentration is to be determined by titration has been added to the burette. The indicator used for such an acid-base titration often depends on the nature of the constituents. Common indicators like methyl violet, bromophenol blue, methyl red, methyl orange, *etc.* were used in these types of reactions. When the titration constituents are a weak acid and a weak base, a pH meter or a conductance meter are used [5].

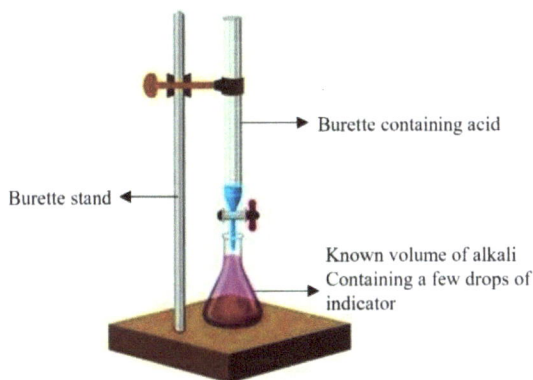

**Fig. (4).** Representing acid base titrations.

## *Acid-Base Indicators*

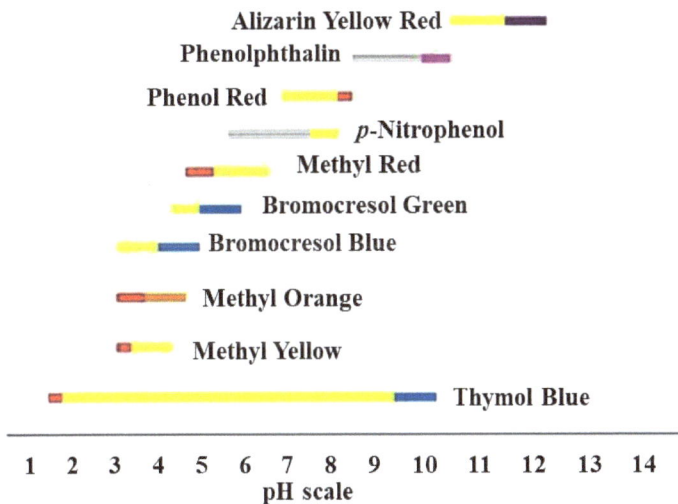

## Redox Titrations

The basic principle of these types of reactions is redox reaction among oxidizing and reducing agents. The oxidizing agent is added to the burette which was rinsed with the same previously. On the other hand, the reducing agent is added to the conical flask, which had been rinsed with distilled water previously. The standard solution is often taken in the conical flask, while the solution whose concentration is to be determined is often taken in the burette as in acid-base titrations. Due to the strong colour of some constituents, some redox titrations do not require an indicator. For instance, in a titration where the oxidizing agent potassium permanganate (paermagnometry) is present, a slight faint persisting pink colour signals the endpoint of the titration, and no particular indicator is required [6]. The titrations named after the reagent used are as follows.

- Permanganate Titrations
- Dichromate Titrations
- Iodimetry and Iodometric Titrations

## *Oxidation-Reduction Indicators*

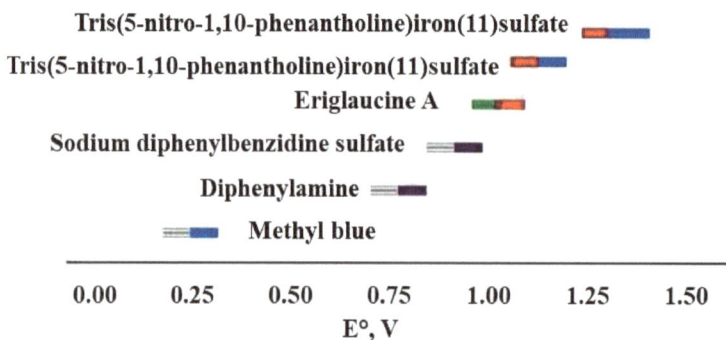

## Precipitation Titration

These titrations are based on the formation of the insoluble precipitates or ions when two reacting substances have been brought into contact (Fig. **5**). A small quantity of potassium chromate ($K_2CrO_4$) solution is added to serve as an indicator.

The first excess of titrant results in the formation of a red silver chromate precipitate which signals the endpoint. Another example includes when silver nitrate solution is added to the ammonium thiocyanate or sodium chloride solution, the formation of silver thiocyanate and silver chloride precipitates takes place.

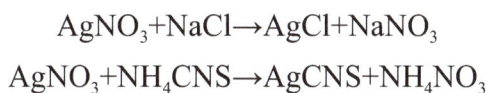

$$AgNO_3 + NaCl \rightarrow AgCl + NaNO_3$$

$$AgNO_3 + NH_4CNS \rightarrow AgCNS + NH_4NO_3$$

**Fig. (5).** Illustrating precipitation titrations.

## Complexometric Titration

These titrations are based on the formation of a complex between the analyte and the titrant. The chelating agent EDTA is very commonly used to titrate metal ions in solution. These titrations generally require specialized indicators that form weaker complexes with the analyte. A common example is Eriochrome Black T and muroxide for the titration of calcium and magnesium ions [7].

## Back Titration

Let us assume that we have an impure solid substance, 'C', weighing 'w' g and we need to calculate the percentage purity of 'C' in the sample. We are also provided with two solutions 'A' and 'B', where the concentration of 'B' is known

(N1) and that of 'A' is unknown.

For the back titration to work, the following conditions are to be satisfied:

- Compounds 'A', 'B' and 'C' should be such that 'A' and 'B' react with each other.
- 'A' and pure 'C' also react with each other, but the impurity present in 'C' does not react with 'A'.
- Also, the product of 'A' and 'C' should not react with 'B'.

Now we take out a certain volume of 'A' in a flask (the equivalents of 'A' taken should be > equivalents of pure 'C' in the sample) and perform a simple titration using 'B'. Let us assume that the volume of 'B' used to be $V_1$ liter.

Equivalents of 'B' reacted with 'A' = $N_1V_1$

Hence Equivalents of 'A' initially = $N_1V_1$

In another flask, we again take some volume of 'A', but now 'C' is added to this flask. The pure part of 'C' reacts with 'A' and excess of 'A' is back titrated with 'B'. Let the volume of 'B' consumed be $V_2$ liter.

The equivalent of 'B' reacted with an excess of 'A' = $N_1V_1$

Hence Equivalents of 'A' initially = $N_1V_1$

Equivalents of pure 'C' = $(N_1V_1 - N_1V_2)$

Let the n-factor of 'C' in its reaction with 'A' be x, then the moles of pure 'C' =

Hence Mass of pure 'C' = $\times$ Molar mass of 'C'.

Hence Percentage purity of 'C' = $\times \times 100$

## *Double Titrations*

The purpose of double titration is to determine the percentage composition of an alkali mixture or an acid mixture. In the present case, we will find the percentage composition of an alkali mixture. Let us consider a solid mixture of NaOH, $Na_2CO_3$ and some inert impurities, weighing 'w' g. We are required to find the % composition of this alkali mixture. We are also given an acid reagent (HCl) of known concentration M1 that can react with the alkali sample.

We first dissolve this mixture in water to make an alkaline solution and then we add two indicators (Indicators are substances that indicate a color change of solution when a reaction gets completed), namely phenolphthalein and methyl orange to the solution. Now, we titrate this alkaline solution with standard HCl.

NaOH is a strong base while $Na_2CO_3$ is a weak base. So, it is obvious that NaOH reacts first with HCl completely and $Na_2CO_3$ reacts only after complete NaOH is neutralized.

$$NaOH + HCl \rightarrow NaCl + H_2O \tag{i}$$

Once NaOH has reacted completely, then $Na_2CO_3$ starts reacting with HCl in two steps, shown as

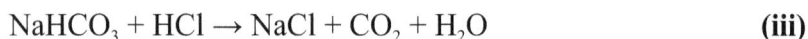

$$Na_2CO_3 + HCl \rightarrow NaHCO_3 + NaCl \tag{ii}$$

$$NaHCO_3 + HCl \rightarrow NaCl + CO_2 + H_2O \tag{iii}$$

It is clear that when we add HCl to the alkaline solution, alkali is neutralized and the pH of the solution decreases. Initially, the pH decrease would be rapid as a strong base (NaOH) is neutralized completely.

When $Na_2CO_3$ is converted to $NaHCO_3$ completely, the solution is still weakly basic due to the presence of $NaHCO_3$ (which is weaker as compared to $Na_2CO_3$). At this point, phenolphthalein changes color since it requires this weakly basic solution to show its color change.

When HCl is further added, the pH again decreases and when all the $NaHCO_3$ reacts to form NaCl, $CO_2$, and $H_2O$, the solution becomes weakly acidic due to the presence of the weak acid ($H_2CO_3$). At this point, methyl orange changes color as it requires this weakly acidic solution to show its color change.

Thus, in general, phenolphthalein shows color change when the solution contains weakly basic $NaHCO_3$ along with other neutral substances, while methyl orange shows color change when the solution has weak acidic $H_2CO_3$ along with other neutral substances.

Let the volume of HCl used up for the first and the second reaction be $V_1$ liter (this is the volume of HCl used from the beginning of the titration up to the point when phenolphthalein shows color change) and the volume of HCl require the third reaction be $V_2$ liter (this is the volume of HCl used from the point where phenolphthalein had changed color up to the point when methyl orange shows color change). Then,

Moles of HCl consumed by $NaHCO_3$ = Moles of $NaHCO_3$ reacted = $M_1V_2$

Moles of $NaHCO_3$ formed from $Na_2CO_3 = M_1V_2$

Moles of $Na_2CO_3$ in the mixture = $M_1V_2$

Mass of $Na_2CO_3$ in the mixture = $M_1V_2 \times 106$

% of $Na_2CO_3$ in the mixture =

Moles of HCl used in the reaction (i) and (ii) = $M_1V_2$

Moles of HCl used in the reaction (ii) = $M_1V_2$

Moles of HCl used in the reaction (i) = $(M_1V_2 - M_1V_2)$

Hence Moles of NaOH = $(M_1V_2 - M_1V_2)$

Mass of NaOH = $(M_1V_2 - M_1V_2) \times 40$

% of NaOH in the mixture =

Here, we have determined the percentage composition of the mixture using the mole concept, as the balanced reactions were available. If we were to solve this by equivalent concept, then the adopted procedure would be.

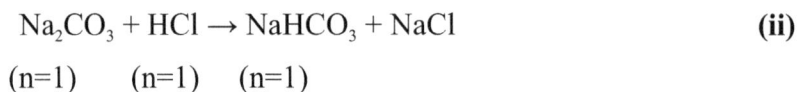

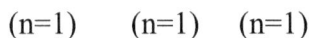

$$NaOH + HCl \rightarrow NaCl + H_2O \qquad \textbf{(i)}$$
$$Na_2CO_3 + HCl \rightarrow NaHCO_3 + NaCl \qquad \textbf{(ii)}$$
$$(n=1) \qquad (n=1) \qquad (n=1)$$

Phenolphthalein shows endpoint after reaction (ii)

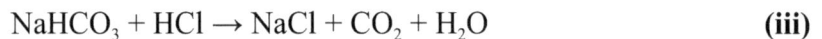

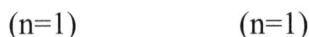

$$NaHCO_3 + HCl \rightarrow NaCl + CO_2 + H_2O \qquad \textbf{(iii)}$$
$$(n=1) \qquad\qquad (n=1)$$

and methyl orange shows the endpoint after reaction (iii).

At phenolphthalein endpoint [8].

**Steps Involved in Titrimetric Analysis**

1. Sampling

2. Preparation of titrant

3. Preparation of standard solution by titration of an accurately known quantity of standard

4. Preparation of sample and measurement of sample

5. Titration with titrant solution

6. Analysis of data

Some Important Rules of Thumb for a Successful Titration are:

- Titrant used for titration should be a standard solution.
- Completion of reaction should be there. The reaction should proceed up to a stable and definite endpoint.
- Equivalence point must be clear.
- The mass of titrant, the volume of sample, and the mass of sample must be known accurately.
- The reaction should be nearly completed to the equivalence point, which means the chemical equilibrium must favor the product.
- The chemical reaction among products must be complete and the rate of reaction should be fast.
- A visual indicator or an instrumental method must be there for detecting the endpoint of the reaction.
- Before use, all the apparatus like measuring cylinders, burettes, and pipettes must be calibrated before use.

## Primary Standard Solution

These are highly purified compounds considered as reference material in all the volumetric methods. These compounds should have properties like high purity, stability towards air, availability, cost-friendly, absence of hydrate water, moderate solubility in the titration medium, a larger molar mass so that error can be minimized.

In this titration, endpoints are shown by using an indicator. Moreover, other methods like the use of visual indicators (change in colour of the reaction mixture). In acid-base reactions, pH indicators such as phenolphthalein have been used, which become pink at certain pH of about 8.2 and methyl orange, which becomes red in acids and yellow in alkali. Although, some titration like redox titrations does not require any indicators are the colour of either reactant or the product is strongly coloured.

For example, potassium permanganate has pink and purple colour used as a titrant in some redox titrations. Such reactions do not need any indicator.

## Methods of Expressing Concentrations in Volumetric Analysis

Standard solution is one which contains a known weight of the reagent in a definite volume of the solution.

Molar solution is one, which contains 1 gm molecular weight of the reagent per liter of solution.

$$\mathbf{M} = \frac{\text{weight}}{\text{M. Wt}} \times \frac{1000}{\text{volume (ml)}}$$

Normal solution is one that contains 1gm equivalent weight per liter of solution.

$$\text{N} = \frac{\text{weight}}{\text{eq. Wt}} \times \frac{1000}{\text{volume (ml)}}$$

Part per million(ppm): Milligrams of solute per liter of solution.

$$\mathbf{ppm} = \frac{\text{weight of solute (mg)}}{\text{Volume of solution (L)}}$$

For titrimetric reaction

$$aA+bB \rightarrow \text{product}$$

At equivalent point:

No. mmol of titrant (A)= No. mmol of titrant (B)

$$N_A \times V_A = N_B \times V_B$$

## Calculation of Equivalent Weights

The definition of equivalent weight depends upon the type of reaction involved in titration. Some examples are given as below:

## Equivalent Weight in Neutralization Reactions

The equivalent weight of acid is a one-gram atom of replaceable hydrogen.

Ex: equivalent weight of $H_2SO_4$ =Mol. Wt. $H_2SO_4$/2

Equivalent weight of $H_3PO_4$ =Mol. Wt. $H_3PO_4$/3

$$\textbf{Eq. wt. of an acid} = \frac{\text{Molecular wt. acid}}{\text{No. of replaceable } H^+}$$

The equivalent weight of Base contains one replaceable hydroxyl group.

Example: equivalent weight of NaOH =Mol. Wt. NaOH/1

$$\textbf{Eq. wt. of a base} = \frac{\text{Molecular wt. base}}{\text{No. of replaceable } OH^-}$$

## Equivalent Weight in Oxidation -reduction Reactions

The equivalent weight of an oxidant or a reductant is the number of electrons of which 1moL of the substance gains or losses in the reaction.

$$\textbf{Eq. wt. of an oxidising and reducing agent} \quad \frac{\text{Molecular wt.}}{\text{No. of electrons lost or gained}}$$

Example:

$$5C_2O_4^= + 2MnO_4^- + 16H^+ \longrightarrow 10CO_2 + 2Mn^{2+} + 8H_2O$$

$$eq.wt.MnO_4^- = \frac{F.wt\ MnO_4^-}{5} \qquad\qquad eq.wt.C_2O_4 = \frac{F.wt\ C_2O_4^=}{2}$$

## Equivalent Weight of Complex Formation and Precipitation Reactions

Here the equivalent weight is the weight of the substance which contains or reacts with 1g. atm of a univalent cation $M^+$ [9, 10].

Ex: When silver nitrate reacts with sodium chloride, to form silver chloride, the equivalent weight of $AgNO_3$ is:

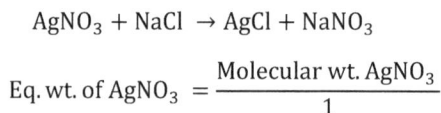

$$AgNO_3 + NaCl \rightarrow AgCl + NaNO_3$$

$$Eq.\,wt.\,of\,AgNO_3 = \frac{Molecular\,wt.\,AgNO_3}{1}$$

## CONSENT FOR PUBLICATION

Not applicable.

## CONFLICT OF INTEREST

The author declares no conflict of interest, financial or otherwise.

## ACKNOWLEDGEMENTS

Declared none.

## REFERENCES

[1]     Whitney, W.D.; Smith, B.E. Titrimetry. In: *The Century Dictionary and Cyclopedia*; The Century Co., **1911**; p. 6504.

[2]     Harris, D.C. Quantitative Chemical Analysis (6th ed.). *Macmillan,* **2003**.

[3]     Compendium for Basal Practice in Biochemistry. Aarhus University, **2008**.

[4]     Titrand. *Science & Technology Dictionary,* **2011**,

[5]     pH measurements with indicators. **2011**.

[6]     Vogel, A.I.; Mendham, J. Vogel's textbook of quantitative chemical analysis (6th ed.). Prentice Hall , **2000**.

[7]     Khopkar, S.M. Basic Concepts of Analytical Chemistry (2nd ed.). New Age International , **1998**; pp. 63-76.

[8]     Kenkel, J. Analytical Chemistry for Technicians *Vol. 1 (3rd ed.). CRC Press,* **2003**, 108-109.

[9]     Kasture, A.V Madhik, K.R.; Wadodkar, S.G.; More, H.N. Pharmaceutical Analysis. *Vol. – 1 (9th ed.) Nirali Prakashan,* **2014**, 3.1-3.9. ISBN 978-81-85790-07-7.

[10]    Fernando, Q M.D. Calculations in analytical chemistry. *New York : Harcourt Brace Jovanovich,* **1982**. ISBN: 0155057103 9780155057104.

# CHAPTER 2

# Recent Advances and Future Perspectives in Volumetric Analysis

**Nidhi Garg**[1,*], **Anju Goyal**[1] and **Payal Das**[1]

[1] *Chitkara College of Pharmacy, Chitkara University, Punjab, India*

**Abstract:** In recent years, analytical methods and their validations have played an important role in the quantification of drugs from their dosage forms or biological samples. The development of various analytical methods with others is very useful for the investigation of the behaviour of drugs or metabolites or impurities and is also a useful tool for sensitive detections. In the 18th century, volumetric analysis, also known as titrimetry, was developed as a control method for determining potash, sulfuric acid, and, later, hypochlorite, all solutions used in textile bleaching in the textile industry. The first methods developed were for practical purposes, to control the 'goodness' (in French, titer) of solutions rather than to determine accurate concentrations. Volumetric analysis is a quantitative analysis used to determine the concentrations of unknown substances. A reaction occurs when the titrant (known solution) is applied to an unknown quantity of analyte (unknown solution) and the concentration of the unknown substance is calculated by knowing the volume of the titrant. Automated titration equipment is used in medical laboratories and hospitals for the same purpose. Apart from these, the process is commonly used in analytical laboratories, as well as the pharmaceutical, food, and petrochemical industries. It is used to calculate the acidity of a sample of vegetable oil in the biodiesel industry. Volumetric analysis has also been used in space science to assess the existence of volatile. Titrimetry is one of the oldest analytical techniques, and it is still commonly used due to its high precision, accuracy, ease of use, and low cost. Titration is therefore seen as a very basic and effective technique that can be used in a number of chemical analysis applications. This chapter describes the development of large-scale industry, recent roles of titrimetry, its applications and future perspectives.

**Keywords:** Analytical methods, Analyte, Applications, Biodiesel industry, Biological samples, Chemical analysis, Dosage forms, Drug analysis, Drug development, Food industry, Future perspectives, Large-scale industry, Medical laboratories, Metabolites, Pharmaceuticals, Space science, Titrant, Vegetable oil, Volumetric analysis.

* **Corresponding Author Nidhi Garg**: Chitkara College of Pharmacy, Chitkara University, Punjab, India;
E-mail: nidhi_garg08@yahoo.co.in

**Anju Goyal & Harish Kumar (Eds.)**
**All rights reserved-© 2022 Bentham Science Publishers**

# INTRODUCTION

Industrialists, scientists, and representatives have been watching the progress on drug production and analytical approaches as a result of technical developments. Bulk drug compounds, intermediate products, drug products, drug formulations, degradation products, impurities, and biological samples, and drug metabolites are vital. Literature survey showed that analytical methods of medicine involve spectroscopy, chromatography, titrimetry, electrochemistry, and capillary electrophoresis. The physical and chemical features of drug compounds can be investigated by these methods [1]. Stability parameters, storage or working conditions, impurity detection, and sensitive quantifications are all possible.

In the 18$^{th}$ century, the textile industry developed volumetric analysis (titrimetry) as a control method for determining sulfuric acid, potash, and later, hypochlorite, both solutions are used in textile bleaching. The first methods were developed for practical reasons, such as controlling the 'goodness' (in French, titer) of solutions rather than determining precise concentrations. Home described a method for testing the 'goodness' of potash in Experiments on Bleaching, 1756, in which nitric acid of a given dilution (1–6) was applied by the teaspoonful to 1 drachm of potash until the effervescence stopped. Potash was not considered suitable for bleaching if less than 12 teaspoonfuls of the acid mixture were needed. Lewis used litmus paper instead of effervescence to signal the end of a reaction in 1767. He weighed the titrant to find out how much he wanted. He used pure potassium carbonate to calculate the acid titer, allowing him to make absolute determinations [2].

Titrimetry is one of the oldest analytical techniques, and it is still commonly used due to its high precision, accuracy, ease of use, and low cost. It is now regarded as one of the key methods of analysis, according to the opinion of the Comite Consultatif pour la Quantite de la Matière (CCQM), *i.e.*, it meets the demands of the highest metrological qualities. Titration is therefore seen as a very basic and effective technique that can be used in a number of chemical analysis applications. Volumetric analysis is a quantitative analysis to determine the concentration or the number of substances involving a chemical reaction. Basic principles take account of an unknown amount of chemical contained in a solution to be analyzed, completion of the reaction by reacting the reagent of unknown concentration with an unknown amount in the presence of an indicator denoting its endpoint. After the completion of the reaction between the solution and the reagent, volumes are studied through titrimetric analysis. The amount of volume and concentration of reagent used in the titration gives the measure of the reagent and the solution. By applying mole fraction of equation exact amount of an unknown chemical in the distinctive volume of solution is determined [3]. A physical chemist might use titration to figure out equilibrium constants, while an analytical chemist might use one to figure out the concentration of one or more components in a sample.

A technique invented by a French chemist, Jean-Baptiste-André Dumas, for determining the proportion of nitrogen combined with other elements in organic compounds is an example of the first process. In a furnace, a weighted sample of the compound is burned under conditions that ensure that all of the nitrogen is converted to elemental nitrogen gas, $N_2$. The nitrogen is transported from the furnace in a stream of carbon dioxide, which is absorbed by a heavy alkali solution, allowing the nitrogen to collect in a graduated tube. The mass of nitrogen can be determined from the amount it takes up under known temperature and pressure conditions, allowing the proportion of nitrogen in the sample to be calculated.

In the study of nitrates, which can be converted into nitric oxide, NO, a gas, a volumetric approach is often used. Carbon dioxide production and consumption are often measured volumetrically during biological processes. The volume shifts that occur as the sample are treated successively with reagents that precisely absorb components such as carbon dioxide, carbon monoxide, oxygen, and others may be used to determine the composition of fuel gases and combustion products [1].

## TYPES OF VOLUMETRIC TITRATIONS

### Acid-Base Titration

Among the various types of titrations, this one is unquestionably the most significant. Acidimetry is the measurement of acid strength using a regular solution of the base. Alkalimetry, on the other hand, is a method of determining the strength of a base using a regular acid solution.

Both titrations certainly play a role in the neutralization of alkali. Furthermore, in an acid-base titration, one of the solutions is an acid and the other a base. Moreover, one is placed in a flask and the other is placed in a burette, from which it is dripped into the flask until the titration reaches its endpoint.

Examples:

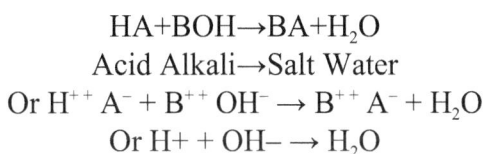

$$HA + BOH \rightarrow BA + H_2O$$
Acid Alkali → Salt Water
$$Or\ H^{++}\ A^- + B^{++}\ OH^- \rightarrow B^{++}\ A^- + H_2O$$
$$Or\ H+ + OH- \rightarrow H_2O$$

This titration is based on the reaction of neutralization between a base or an acidic and analyte. Furthermore, in this type, a reagent is mixed with the sample solution until it reaches the required pH level. This type of titration majorly depends on the track change in pH or a pH meter.

## Redox Titrations

This titration can be described as an oxidation-reduction reaction. The chemical reaction in this titration is a transfer of electrons between reacting ions in aqueous solutions. One solution is a reducing agent, whereas the other is an oxidizing agent in redox titration.

Examples:

Some popular examples of these titrations are as follows:

### *Permanganate Titrations*

The use of potassium permanganate is an oxidizing agent. Its maintenance takes place with the use of dilute sulphuric acid.

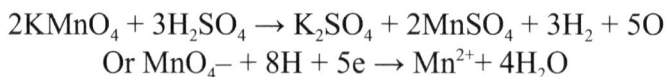

$$2KMnO_4 + 3H_2SO_4 \rightarrow K_2SO_4 + 2MnSO_4 + 3H_2 + 5O$$
$$Or\ MnO_4- + 8H + 5e \rightarrow Mn^{2+} + 4H_2O$$

Before the endpoint, this solution is colorless. Furthermore, potassium permanganate is used to measure oxalic acid, ferrous salts, hydrogen peroxide, oxalates, and several other compounds.

### *Dichromate Titrations*

These are certainly using potassium dichromate as an oxidizing agent in an acidic medium. Moreover, the maintenance of the acidic medium takes place by the use of dilute sulphuric acid.

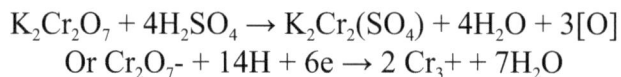

$$K_2Cr_2O_7 + 4H_2SO_4 \rightarrow K_2Cr_2(SO_4) + 4H_2O + 3[O]$$
$$Or\ Cr_2O_7- + 14H + 6e \rightarrow 2\ Cr_3++ 7H_2O$$

### *Iodometric and Iodometric Titrations*

Furthermore, during these titrations, the reduction of free iodine to iodide ions and oxidation of iodide ions to free occurs.

$$I_2 + 2e \rightarrow 2I^- \ldots\ldots\ldots\ldots \text{(reduction)}$$
$$2I^- + 2e \rightarrow 2e \ldots\ldots\ldots\ldots \text{(oxidation)}$$

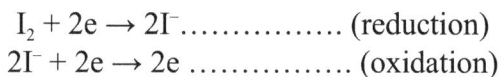

## Precipitation Titrations

This titration is based on the precipitate formation. In precipitation titration, two reacting substances are brought into contact.

Example: When the use of the solution of silver nitrate takes place to a solution of ammonium thiocyanate or sodium chloride. It reacts and forms a white precipitate of silver thiocyanate or silver chloride.

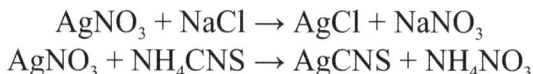

$$AgNO_3 + NaCl \rightarrow AgCl + NaNO_3$$
$$AgNO_3 + NH_4CNS \rightarrow AgCNS + NH_4NO_3$$

## Complexometric Titrations

The formation of an undissociated complex is the most significant result of this titration. It's much more than the titrations of precipitation.

Example:

$$Hg^{2+} + 2SCN- \rightarrow Hg(SCN)_2$$
$$Ag+ + 2CN- \rightarrow [Ag(CN)_2]^-$$

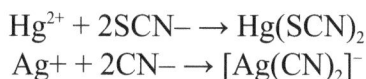

Ethylenediaminetetraacetic acid *i.e.*, EDTA is certainly an important reagent that forms complexes with metals.

## PROCEDURE FOR VOLUMETRIC ANALYSIS

1. A standard titration begins with a measured burette or pipette containing the titrant, a small amount of indicator placed underneath a calibrated burette or pipette containing the titrant.

2. The solution that needs to be analyzed must have a reliable weighed sample of +/- 0.0001g of the material to be analyzed.

3. It is also essential to choose the right kind of material to analyze since the wrong choice of titrant would give us the wrong results. Substances that react rapidly and completely to produce a complete solution are chosen.

4. Small quantities of titrant are applied to the analyte and indicator until the indicator changes color in response to the titrant saturation threshold, signaling that the titration has reached its endpoint.

5. The titration must be continued until the reaction is finished and the amount of reactant added is exactly the amount required to complete the reaction.

6. Since molarity is a common metric for calculating the number of moles present in a solution, determining the correct volume of the standard solution is also crucial.

7. Focusing on the desired endpoint, a single drop or less than a drop of the titrant would make a difference between a permanent and temporary shift in the indicator.

8. We should weigh and dissolve the reagent into a solution so that it is in a specified volume in a volumetric flask if the reagent or reactant we use is to be made into a standard solution.

9. Calculate the normality of the titrant from the final reading using the equation:

$$N_1V_1 = N_2V_2$$

10. Finally, after calculating normality, then calculate the amount of given substance in the whole solution using the equation:

Mass= EquivalentWeight x Normality x Volume100/frac {Equivalent Weight x Normality x Volume} {100} 100 Equivalent weight x Normality x Volume

## BASIC PRINCIPLES OF VOLUMETRIC ANALYSIS

1. A significant amount of chemicals is present in the solution to be analyzed.

2. To display the endpoint, a reagent of unknown concentration reacts with a chemical of an unknown amount in the presence of an indicator (usually phenolphthalein). It is the point that indicates the completion of the reaction.

3. The volumes are measured by titration, which completes the reaction between the solution and the reagent.

4. The amount of reagent and solution is shown by the volume and concentration of reagent used in the titration.

5. The amount of unknown chemicals in the specific volume of solution is determined by the mole fraction of the equation.

When the reaction reaches its endpoint, the amount of reactant absorbed is determined and used to determine the volumetric analysis of the analyte using the formula below.

$$Ca = Ct \; Vt \; M \, / \, Va$$

Where,

Ca in molarity is the analyte concentration.

Ct typically in molarity is the titrant concentration.

V in liters is the volume of titrant which is used.

M is the mole ratio of the analyte and reactant from the balanced equation.

V in liters is the volume of the analyte.

A constant pH in the reaction necessitates several non-acid-base titrations. To retain pH value, a buffer solution can be added to the titration chamber. Quality management is a checkpoint in the pharmaceutical industry for both raw materials and finished products. In quality control, the incoming material is first tested using the titrimetric process. By conducting qualitative and quantitative analysis, the purity or quality of the substance is verified. Even though volumetric titration is not a completely accurate method since the outcome is influenced by the titration equipment, temperature, and other laboratory conditions. To obtain accurate and transparent results, it is recommended that the titer of volumetric solutions be determined. The method result is influenced by many factors, such as measuring method instrument (instrument error/burette abrasion), Electrodes (electrode error/electrode alteration), Handling (*e.g.* dilution preparation), Balance (weighing error), Temperature change in the volumetric solution as a result of environmental influences such as oxygen, carbon dioxide, and microorganisms.

## IMPORTANCE OF VOLUMETRIC ANALYSIS

Chemists care about the quantitative interaction between two reacting solutions. In a chemical analysis involving solutions up to some point, solid precipitates of chemical reactions between such solutions were dried, separated, and massed.

Gravimetric analysis is the name of the technique. It is used to determine mass relationships in quantitative experiments. While the method is useful, it is not always practical. Separating and measuring the mass of reaction products when they are in solutions is challenging and, in many cases, a waste of time and materials.

The volumetric analysis appears to be a better and faster technique, especially when acids and bases are involved. For better quantitative data, they can be titrated against one another.

In high school, college, and university chemistry laboratories, volumetric analysis is used to assess the concentrations of unknown substances Table **1**. A reaction occurs when the titrant (known solution) is applied to an unknown quantity of analyte (unknown solution). The student will calculate the concentration of the unknown substance by knowing the volume of the titrant.

**Table 1. Recent studies on drugs were analyzed by volumetric analysis.**

| S. No. | Drugs | Outcome: | Reference |
|---|---|---|---|
| 1. | Levofloxacin hemihydrate | As the drug contain both hydrophilic and hydrophobic group, the reaction shows variable kind of molecular interaction. | Lomesh *et al.*, 2019 [7] |
| 2. | Norfloxacin | The study shows an increase in the value of partial molar volume with the increase in surfactant in the absence of norfloxacin and a decrease in value with the addition of surfactant and presence of norfloxacin. | Sohail *et al.*, 2020 [8] |
| 3. | Domiphen bromide | The ion-ion/polar interactions between DB and amino acids/dipeptides are dominant. | Yan *et al.*, 2016 [9] |
| 4. | Sodium ibuprofen | Volume and compression data suggest the dominance of hydrophilic–hydrophilic interactions in amino acids–drug mixtures. | Sharma *et al*, 2016 [10] |
| 5. | Isoniazid | Ion-solvent and solvent-solvent interactions increase with the increase in the concentration of the drug isoniazid while decreasing with the increase in temperature in an aqueous carbohydrate solution. Due to the presence of great attractive force within the solvent molecules, the drug shows extra density in aqueous-arabinose than in aqueous-xylose. | Naina *et al.*, 2019 [11] |
| 6. | Semicarbazide hydrochloride | Elucidates interactions of semicarbazide HCl with carbohydrates in aq. media. | Chand *et al.*, 2020 [12] |
| 7. | Doxycycline Hyclate | The value of the apparent molar volume increases with the increase in drug concentration. | Lomesh, *et al.*, 2019 [13] |

Automated titration equipment is used in medical laboratories and hospitals for the same purpose [4]. Aside from these, the process is commonly used in analytical laboratories, as well as the pharmaceutical, food, and petrochemical industries. It is used to calculate the acidity of a sample of vegetable oil in the biodiesel industry, for example.

Scientists can determine the exact amount of base necessary to neutralize a sample of vegetable oil by knowing how much base to add to neutralize the entire amount [5].

Volumetric analysis has also been used in space science to assess the existence of volatile components in the ejecta flow of crater cavity volume, in ecological studies to determine the relationship between brain structure and sensory ecology of aquatic animals – teleost fishes, in specialty metal application and a number of other fields [6].

It is an area of science that no man can do without if he wants to live a healthy and good life.

## APPLICATIONS AND FUTURE PERSPECTIVES OF VOLUMETRIC ANALYSIS

1. Volumetric analysis can be used to analyze glioblastomas (A malignant tumor affecting the brain or spine).

2. The use of volumetric analysis of metformin hydrochloride in water.

3. The use of volumetric analysis to check the interaction between the drug isoniazid in aqueous D-Xylose/L-arabinose solution.

4. The use of volumetric analysis for alteration in brain perfusion.

5. Application of volumetric analysis on the caudate nucleus in Alzheimer's disease.

6. Volumetric analysis on the brain stem and cerebellum on Machado joseph disease (inherited neurodegenerative disease).

7. The use of volumetric analysis for the interaction of L-histidine in aqueous ionic liquid at different temperature.

8. Application of volumetric analysis of binary liquid mixture containing ethyl---hydroxy benzoate and alkanols.

9. Application of volumetric analysis to check the effect of fruit and milk sugar in aq. Solution of the drug diphenhydramine- hydrochloride.

10. The use of volumetric analysis to check the aggregation of bio surfactants in aqueous solution galactose and lactose.

11. To verify the effect of sodium chloride and myo-inositol in aq. Solution of diphenhydramine hydrochloride drug.

12. Volumetric analysis of the hypothalamus in Huntington disease with the help of 3T MRI.

13. The use of volumetric analysis of betaine hydrochloride drug in aq. Uracil sol.

14. The use of Volumetric analysis of structural isomers D (+) glucose and D(-) fructose in aq. dipotassium oxalate solution.

15. Volumetric analysis of naproxen in aqueous solution of choline based deep eutectic solvents.

16. The use of volumetric analysis in the substantia innominate of parkinson's patients.

## CONCLUSION

Volumetric analysis, any method of quantitative chemical analysis in which the amount of a substance is determined by measuring the volume that it occupies or, in broader usage, the volume of a second substance that combines with the first in known proportions, more correctly called titrimetric analysis. This chapter includes types of volumetric titration like acid-base titrations, redox titrations, iodometry and iodimetry titrations, precipitation titrations and complexometric titrations. The basic principle involved in the volumetric analysis is the detection of the amount of unknown substance present in the solution using an indicator. The use of volumetric titrations in calculating the acidity of vegetable oil, application in space science, alteration in brain perfusion, *etc.* has been explained in the current chapter.

## CONSENT FOR PUBLICATION

Not applicable.

## CONFLICT OF INTEREST

The author declares no conflict of interest, financial or otherwise.

## ACKNOWLEDGEMENTS

Declared none.

# REFERENCES

[1]     Bonfilio, R.B.D.A.M.; De Araujo, M.B.; Salgado, H.R.N. Recent applications of analytical techniques for quantitative pharmaceutical analysis: a review. *WSEAS Trans. Biol.,* **2010**, *7*(4), 316-338.

[2]     Worsfold, P.; Poole, C.; Townshend, A.; Miro, M. History of Analytical Science.*Encyclopedia of Analytical Science,* 3rd ed; Elsevier: UK, **2019**, pp. 399-410.

[3]     Michałowska-Kaczmarczyk, A.M.; Spórna-Kucab, A.; Michałowski, T. Principles of Titrimetric Analyses According to Generalized Approach to Electrolytic Systems (GATES). In: *Advances in Titration Techniques, Vu Dang Hoang, Intech Open*; , **2017**.
[http://dx.doi.org/10.5772/intechopen.69248]

[4]     Joosten, A.; Delaporte, A.; Alexander, B.; Su, F.; Creteur, J.; Vincent, J.L.; Cannesson, M.; Rinehart, J. Automated titration of vasopressor infusion using a closed-loop controller: *in vivo* feasibility study using a swine model. *Anesthesiology,* **2019**, *130*(3), 394-403.
[http://dx.doi.org/10.1097/ALN.0000000000002581] [PMID: 30608239]

[5]     Lee, S.; Posarac, D.; Ellis, N. Lee, S/; Posarac, D/; Ellis, N. Process simulation and economic analysis of biodiesel production processes using fresh and waste vegetable oil and supercritical methanol. *Chem. Eng. Res. Des.,* **2011**, *89*(12), 2626-2642.
[http://dx.doi.org/10.1016/j.cherd.2011.05.011]

[6]     Vervoort, N.; Goossens, K.; Baeten, M.; Chen, Q. *Recent advances in analytical techniques for high throughput experimentation.,* **2021**, *2*(3-4), 109-127.
[http://dx.doi.org/10.1002/ansa.202000155]

[7]     Lomesh, S.K.; Bala, M.; Nathan, V. Physicochemical approach to study the solute-solute and solute-solvent interactions of drug Levofloxacin hemihydrate in aqueous sorbitol solutions at different temperatures: Volumetric, acoustic and conductance studies. *J. Mol. Liq.,* **2019**, *283*(5), 133-146.
[http://dx.doi.org/10.1016/j.molliq.2019.03.055]

[8]     Sohail, M.; Rahman, H.M.; Asghar, M.N.; Shaukat, S. Volumetric, acoustic, electrochemical and spectroscopic investigation of norfloxacin–ionic surfactant interactions. *J. Mol. Liq.,* **2020**, *318*, 114179.
[http://dx.doi.org/10.1016/j.molliq.2020.114179]

[9]     Yan, Z.; Wen, X.; Kang, Y.; Chu, W. Intermolecular interactions of α-amino acids and glycyl dipeptides with the drug domiphen bromide in aqueous solutions analyzed by volumetric and UV–vis spectroscopy methods. *J. Chem. Thermodyn.,* **2016**, *101*, 300-307.
[http://dx.doi.org/10.1016/j.jct.2016.06.018]

[10]    Sharma, S.K.; Singh, G.; Kumar, H.; Kataria, R. Solvation behavior of some amino acids in aqueous solutions of non-steroidal anti-inflammatory drug sodium ibuprofen at different temperatures analysed by volumetric and acoustic methods. *J. Chem. Thermodyn.,* **2016**, *98*, 214-230.
[http://dx.doi.org/10.1016/j.jct.2016.03.016]

[11]    Naina, A.K. Volumetric, acoustic and viscometric studies of solute-solute and solute-solvent interactions of isoniazid in aqueous-glucose/sucrose solutions at temperatures from 293.15 K to 318.15 K. *J. Chem. Thermodyn.,* **2019**, *133*, 123-134.
[http://dx.doi.org/10.1016/j.jct.2019.01.024]

[12]    Chand, D.; Nain, A.K. Molecular interactions of drug semicarbazide hydrochloride in aqueous---xylose/L-arabinose solutions at different temperatures: Volumetric, acoustic and viscometric study. *J. Chem. Thermodyn.,* **2020**, *146*, 106106.
[http://dx.doi.org/10.1016/j.jct.2020.106106]

[13]    Lomesh, S.K.; Nathan, V.; Bala, M.; Thakur, P. Volumetric and acoustic methods for investigating molecular interactions of antibiotic drug doxycycline hyclate in water and in aqueous solution of sodium chloride and potassium chloride at different temperatures (293.15–313.15). *J. Mol. Liq.,* **2019**, *284*, 241-251.
[http://dx.doi.org/10.1016/j.molliq.2019.04.006]

# CHAPTER 3

# Aqueous Acid-Base Titrations

**Astha Sharma[1,\*], Monika Gupta[2] and Anju Goyal[3]**

[1] *Laureate Institute of Pharmacy, Kangra, Himachal Pradesh, India*

[2] *A.S.B.A.S.J.S. Memorial College of Pharmacy, Bela Punjab, India*

[3] *Chitkara College of Pharmacy, Chitkara University, Punjab, India*

**Abstract:** Titration is the quantitative addition of a solution of known concentration to a solution of unknown concentration until the reaction between them is complete to determine the concentration of the second solution. An acid-base titration is the quantitative determination of the concentration of an acid or a base. Titration of an acid with a base requires that the pH, or relative concentrations of the two reactants, be monitored. pH can be assessed by litmus paper or by indicators, for example, phenolphthalein, but these methods lack precision. Typically, pH measurement in the laboratory is done by measuring the cell potential of that sample in reference to a standard hydrogen electrode. The endpoint or equivalent points can be determined by the titration curve, which is a plot of the pH of an acidic (or basic) solution that acts as a function of the amount of base (or acid) added.

**Keywords:** Acid, Base, Neutralization, pH, Titration curve.

## INTRODUCTION

Acid-Base titrations or neutralization titrations are usually used to find the appropriate amount of a known acidic or basic substance through acid-base reactions. The analyte (titrant) is the solution with unknown molarity. The reagent (titrant) is the solution with known molarity that will react with the analyte [1]. The reactions in which $H_3O^+$ in solution is titrated by $OH^-$ and this can be applied regardless of strong acid, strong base, weak acids, weak bases, salt of the weak acid, salt of the weak base are in the titrations reactions [2].

$$HCl + NaOH \longrightarrow NaCl + H_2O$$
$$HCl + NH_4OH \longrightarrow NH_4Cl + H_2O$$

\* **Corresponding author Astha Sharma:** Laureate Institute of Pharmacy, Kangra, Himachal Pradesh, India; E-mail: astha110sharma@gmail.com

**Anju Goyal & Harish Kumar (Eds.)**
**All rights reserved-© 2022 Bentham Science Publishers**

The analyte is prepared by dissolving the substance being studied into a solution. The solution is usually placed in a flask for titration. A small amount of indicator is then added into the flask along with the analyte. The reagent is usually placed in a burette and slowly added to the analyte and indicator mixture. The amount of reagent used is recorded when the indicator causes a change in the colour of the solution [3].

Some titrations require the solution to be boiled due to the $CO_2$ created from the acid-base reaction. The $CO_2$ forms carbonic acid ($H_2CO_3$) when dissolved in water that then acts as a buffer, reducing the accuracy of data. After boiling water, most of the $CO_2$ will be removed from the solution allowing the solution to be titrated to a more accurate endpoint. The endpoint is the point where all of the analytes have reacted with the reagent [4].

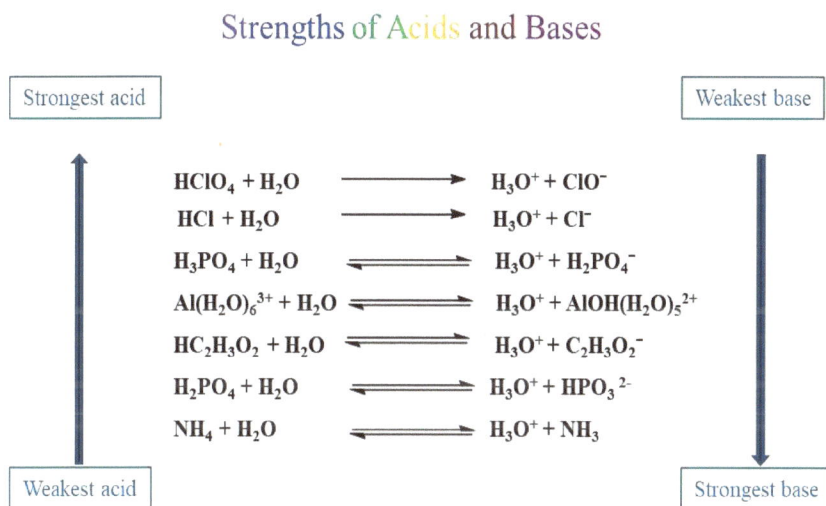

## Strengths of Acids and Bases

| Strongest acid | | Weakest base |
|---|---|---|

$HClO_4 + H_2O \longrightarrow H_3O^+ + ClO^-$

$HCl + H_2O \longrightarrow H_3O^+ + Cl^-$

$H_3PO_4 + H_2O \rightleftharpoons H_3O^+ + H_2PO_4^-$

$Al(H_2O)_6^{3+} + H_2O \rightleftharpoons H_3O^+ + AlOH(H_2O)_5^{2+}$

$HC_2H_3O_2 + H_2O \rightleftharpoons H_3O^+ + C_2H_3O_2^-$

$H_2PO_4 + H_2O \rightleftharpoons H_3O^+ + HPO_3^{2-}$

$NH_4 + H_2O \rightleftharpoons H_3O^+ + NH_3$

| Weakest acid | | Strongest base |
|---|---|---|

# THEORIES OF ACIDS AND BASES

## Arrhenius Concept

According to the Arrhenius concept of acids and bases, an acid is a substance that, when dissolved in water, increases the concentration of hydronium ion ($H_3O^+$). Remember, however, that the aqueous hydrogen ion is chemically bonded to water, that is, $H_3O^+$. A base, in the Arrhenius concept, is a substance that, when dissolved in water, increases the concentration of hydroxide ion, $OH^-$ [5].

## Bronsted-Lowry

Bronsted-Lowry Acid can donate a proton. Bronsted-Lowry Base can accept a proton. Must contain a non-bonding pair of electrons. The conjugate base of an acid is the species remaining after the acid has lost a proton. Conjugate acid is the species formed after the base has accepted a proton. Water is the conjugate base of $H_3O^+$ and $Cl^-$ is the conjugate base of HCl [6].

$$HCl + H_2O \leftrightarrow H_3O^+ + Cl^-$$

## Lewis Acids

Bases can donate a pair of electrons and acids can accept a pair of electrons. Covalent bond is formed. Many Lewis Acids do not contain hydrogen [7].

## Law of Mass Action

This law was proposed by Goldberg and wage in 1867. This was expressed by the rate of a chemical reaction being proportional to the active masses of the reacting substances [8].

In dilute solutions, the ideal state active mass may be represented by the concentration of reacting substance.

Now, let us consider a homogeneous reversible reaction

$$A + B \rightleftharpoons C + D$$

According to the law of mass action

$$V_f = K_1 [A][B]$$
$$V_b = K_2 [C][D]$$

Where $V_f$ = velocity of the forward reaction

$V_b$ = velocity of the backward reaction

[A], [B], [C], [D] = Molar concentration of A, B, C and D respectively, $K_1$ and $K_2$ are constant.

At equilibrium

$$V_f = V_b$$

$$K_1[A] [B] = K_2[C] [D]$$
$$K_1 = [C] [D]$$
$$K_2 = [A] [B]$$

Since $K_1$ and $K_2$ are constants $K_1$ fraction must also be a constant.

**Note:** When equilibrium is reached in a reversible reaction, at constant temperature, the product of the molecular concentration of RHS is divided by the product of the molecular concentration of reactant.

Hence,

$$K = \frac{[C][D]}{[A][B]}$$

Where k is the equilibrium constant of the reaction.

In extension, the equilibrium constant for the general reversible reaction

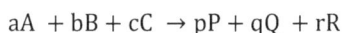

$$aA + bB + cC \rightarrow pP + qQ + rR$$

$$K = \frac{[P]^p [Q]^q [R]^r}{[A]^a [B]^b [C]^c}$$

Where a, b, c and p, q, r are the numbers of molecules of reacting species.

## Ion Product Constant of Water

Aqueous solutions contain small concentrations of hydronium and hydroxide ions as a result of the dissociation reaction.

$$2H_2O \rightleftharpoons H_3O^+ + OH^-$$

- The concentration of water in dilute aqueous solutions is enormous, however, when compared with the concentration of hydronium and hydroxide ions [9].
- Where the new constant *Kw is given a special name, the **ion-product constant for** water.*
- At 25°C, the ion-product constant for water is $1.008 \times 10^{-14}$. For convenience, we use the approximation that at room temperature $Kw = 1.00 \times 10^{-14}$

## COMMON ION EFFECT

- The common-ion effect is a mass-action effect predicted from Le-Châtelier's principle.
- The solubility of an ionic precipitate decreases when a soluble compound containing one of the ions of the precipitate is added to the solution. This behavior is called the **common ion effect.**
- Most, but not all, sparingly soluble salts are essentially **completely dissociated** in saturated aqueous solution [10].

$$Ba(IO_3)_2(s) \rightleftharpoons Ba^{2+}(aq) + 2IO_3^-(aq)$$

- If we add $Ba(NO_3)_2$, which is a strong electrolyte, in the above solution as $[Ba^{2+}]$ increases and according to Le Châtelier's principle, solubility of $Ba(IO_3)_2$ will decrease so as to relieve the stress applied to the system.

## ACID BASE EQUILIBRIUM

Considered a reversible chemical reaction

$$aA + bB \rightleftharpoons cC + dD$$

From the law of mass action

$$K = \frac{[C]^c [D]^d}{[A]^a [B]^b}$$

Now, let us consider the dissociation of weak acid in an aqueous solution

$$HA \rightleftharpoons H^+ + A^-$$

Where $H^+$ represented the hydrated proton

The dissociation constant for weak acid can be denoted as $K_a$ and its value can be represented as:

$$K_a = \frac{[H^+][A^-]}{[HA]}$$

In this water act as a second conjugate acid-base pair

Considered the dissociate of weak base

$$B + H_2O \rightleftharpoons BH^+ + OH^-$$

The value of dissociation constant kb is

$$K_a = \frac{[BH^+][OH^-]}{[B]}$$

In this equation, the water is taken to be constant and it is absorbed in the value of Kb

Water itself is capable of dissociation.

$$H_2O \rightleftharpoons H^+ + OH^-$$

The equilibrium constant

$$K = \frac{[H^+][OH^-]}{[H_2O]}$$

The water is usually considered to be constant

$$Kw = [H^+][OH^-]$$

Kw = ion product of water

Kw is important in order to derive the relationship between Ka and Kb

$$K_a = \frac{[H^+][A^-]}{[HA]}$$

$$K_b = \frac{[HA][OH^-]}{[A]}$$

Multiply these equations

$$K_a \cdot K_b = [H^+] [OH^-]$$

$$K_w = K_a \cdot K_b$$

# INDICATOR

A useful indicator has a strong colour that changes quickly near its pKa. These traits are desirable, so only a small amount of an indicator is needed. If a large amount of indicator is used, the indicator will affect the final pH, lowering the accuracy of the experiment. The indicator should also have a pKa value near the pH of the titration's endpoint. For example, an analyte that is a weak base would require an indicator with a pKa less than 7. Choosing an indicator with a pKa near the endpoint's pH will also reduce error because the colour change occurs sharply during the endpoint where the pH spikes, giving a more precise endpoint. Methyl orange as the indicator colour changes would occur all throughout the region highlighted in pink. The data obtained would be hard to determine due to the large range of colour change and inaccurate as the colour change does not even lie with the endpoint region. Phenolphthalein, on the other hand, changes colour rapidly near the endpoint allowing for more accurate data to be gathered [11, 12].

## Neutralisation Indicator

The substances exhibit different colours at various values of pH. Indicators are weak acid or weak base, which have a different colour in their conjugate acid and base forms [13]. An indicator is a substance that is used to determine the endpoint in a titration. In acid-base titrations, organic substances (weak acids or weak bases) are generally used as indicators. They change their colour within a certain pH range. The colour change and the pH range of some common indicators are tabulated below.

### *Ostwald Theory*

First theory to explain the behaviour of indicators was put forth by W. Ostwald [14]. According to this theory, the dissociated indicator acid or base has a colour different than its ions.

For Acid Indicator

Equivalence can be written as

$$HIn \rightleftharpoons H^+ + In^-$$

By applying the law of mass action

$$KIn_a = \frac{[H^+]\,[In^-]}{[HIn]}$$

$$\log[H^+] = \log KIn_a + \frac{\log[HIn]}{[In^-]}$$

Take negative log both sides

$$-\log[H^+] = -\log KIn_a - \frac{\log[HIn]}{[In^-]}$$

$$pH = pKIn_a + \frac{\log[HIn]}{[In^-]}$$

$KIn_a$ is the dissociation constant of indicator

For base indicator

$$InOH \rightleftharpoons In + OH^-$$

$$KIn_b = \frac{[In^+] + [OH^-]}{[InOH]}$$

$$[OH^-] = \frac{KIn_b + [InOH]}{[In^+]}$$

Now, $\qquad Kw = [H^+]\,[OH^-]$

Substituting this value in the above equation

$$[H]^+ = \frac{Kw\,[In^+]}{KIn_b\,[InOH]}$$

Taking log

$$pH = pKw - pKIn_b - \log \frac{[InOH]}{[In^+]}$$

The colour change will be affected by the $H^+$ concentration and as this change is gradual, the colour change of indicator is also a gradual one.

In order to detect the colour, change the ratio must be at least 1:10.

## Mixed Indicators

The pH ranges are very narrow and colour change over the range must be very sharp. This is possible with ordinary acid-base indicators. The result may be achieved by the use of a suitable mixture of indicator [15]. These are generally selected, so the value is closer together and overlapped colours are complementary at an intermediate pH value.

Example: - A mixture of equal parts of neutral red and methylene blue gives a sharp colour change from violet-blue to green.

A mixture of 3 parts of phenolphthalein and 1 part of alpha-naphtholphthalein passes from pale rose to violet at pH 8.9.

Phenolphthalein:

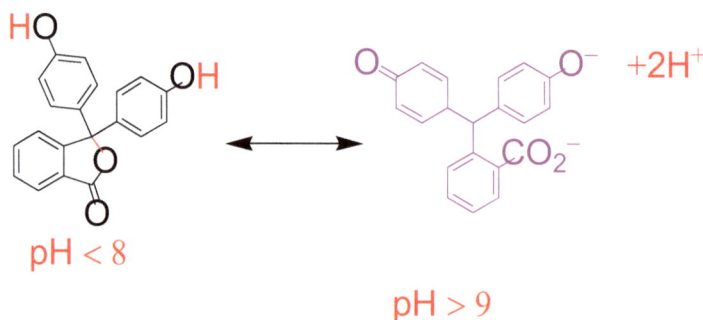

Methyl red:

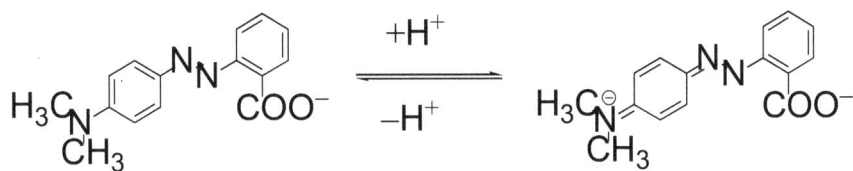

## Universal Indicator

By mixing certain indicators, the colour change may occur to extend over a considerable portion of pH range. Such mixtures are usually called multiple range indicators [16]. They are not suitable for quantitative titration but may be employed to determine the approximate pH of a solution by the colorimetric method.

Example: Dissolve 0.1g of phenolphthalein, 0.2g of methyl red, 0.3g methyl yellow and 0.4g of bromothymol in 1000ml of absolute alcohol and add sufficient sodium hydroxide solution until the color is yellow.

| Colour | pH |
|--------|----|
| Red | 2 |
| Orange | 4 |
| Yellow | 6 |
| Green | 8 |
| Blue | 10 |

## NEUTRALISATION CURVES

The mechanism of neutralisation processes can be understood by studying the changes in the hydrogen ion concentration during the course of the appropriate titration. The change in pH in the neighbourhood of the equivalence point is of the greatest importance, as it enables an indicator to be selected, which will give the smallest titration error [17]. The curve obtained by plotting pH as the ordinate against the percentage of acid neutralised (or the number of mL of alkali added) as abscissa is known as the neutralisation (or, more generally, the titration) curve. This may be evaluated experimentally by the determination of the pH at various stages during the titration by a potentiometric method or it may be calculated from theoretical principles.

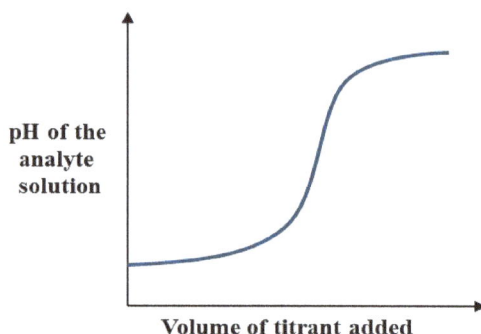

pH of the analyte solution

Volume of titrant added

**1.** Titration of a strong acid with a strong base

**2.** Titration of a weak acid with a strong base

**3.** Titration of a strong acid with a weak base

**4.** Titration of a weak base with a weak acid

### Titration of a Strong Acid with a Strong Base

For this calculation, it is assumed that both the acid and the base are completely dissociated and the activity coefficients of the ions are unity in order to obtain the pH values during the course of the neutralization of the strong acid and the strong base, or vice versa, at the laboratory temperature. For simplicity of calculation consider the titration of 100 mL of 1 M hydrochloric acid with 1 M sodium hydroxide solution [18]. The pH of 1 M hydrochloric acid is O. When 50 mL of the 1 M base has been added, 50 mL of unneutralized 1 M acid will be present in a total volume of 150mL.

$[H^+]$ will therefore be 50 x 11150 = 3.33 x IO-', or pH = 0.48

For 75 mL of base, $[H^+ '1 = 25 \times 11175 = 1.43$ x $10^{-1}$ -',        pH = 0.84

For 90 mL of base, $[H^+ '] = 10 \times 11190 = 5.27$  x $10^{-2}$ -',     pH = 1.3

For 99 mL of base, $[H^+ '] =1 \times 1/199 = 5.03$ x $10^{-3}$ -~,                          pH=2.3

For 99.9 mL of base, $[H^+ = 0.1 \times 11199.9 = 5.01 \times 10^{-4}$ -~, pH = 3.3

Upon addition of 100 mL of base, the pH will change sharply to 7, *i.e.* the theoretical equivalence point. The resulting solution is simply sodium chloride.

Any sodium hydroxide added beyond this will be in excess of that needed for neutralisation.

With 100.1 mL of base, [OH⁻] = 0.1/200.1 = 5.00 x pOH = 3.3 and pH = 10.7

With 101 mL of base, [OH⁻] = 11201 = 5.00 x pOH = 2.3, and pH = 11.7

These results show that as the titration proceeds, initially, the pH rises slowly, but between the addition of 99.9 and 100.1 mL of alkali, the pH of the solution rises from 3.3 to 10.7, i.e. in the vicinity of the equivalence point, the rate of change of pH of the solution is very rapid. The complete results, up to the addition of 200 mL of alkali; this also includes the figures for 0.1 M and 0.01 M solutions of acid and base respectively [19]. The additions of alkali have been extended in all three cases to 200 mL; it is evident that the range from 200 to 100 mL and beyond represents the reverse titration of 100 mL of alkali with the acid in the presence of the non-hydrolysed sodium chloride solution. The data in the table are presented graphically.

In quantitative analysis, it is the changes of pH near the equivalence point which are of special interest. This part also indicated the colour-change intervals of some of the common indicators. With 1 M solutions, it is evident that any indicator with an effective range between pH 3 and 10.5 may be used. The colour change will be sharp and the titration error negligible. With 0.1 M solutions, the ideal pH range for an indicator is limited to 4.5-9.5. Methyl orange will exist chiefly in the alkaline form when 99.8 mL of alkali have been added, and the titration error will be 0.2 percent, which is negligibly small for most practical purposes; it is, therefore, advisable to add sodium hydroxide solution until the indicator is present completely in the alkaline form. The titration error is also negligibly small with phenolphthalein. With 0.01M solutions, the ideal pH range is still further limited to 5.5-8.5; such indicators as methyl red, bromothymol blue, or phenol red will be suitable. The titration error for methyl orange will be 1-2 percent. The above considerations apply to solutions that do not contain carbon dioxide.

**Titration of a Weak Acid with a Strong Base**

The neutralization of 100 mL of 0.1 M acetic acid (ethanolic acid) with 0.1 M sodium hydroxide solution will be considered here; other concentrations can be treated similarly (Fig. **1**) [20].

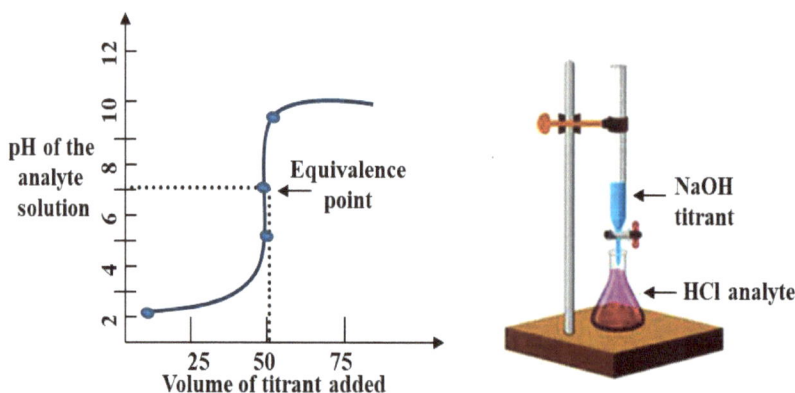

**Fig. (1).** Titration of weak acid with strong base along with graph.

For other concentrations, we may employ the approximate Mass Action expression:

$$[H'] \times [CH_3 COO^-]/[CH_3 COOH] = Ka \qquad (1)$$
$$Or\ [H'] = [CH_3 COOH] \times Ka/[CH_3COO^-]$$
$$Or\ pH = \log [Salt]/[Acid] + pKa$$

The concentration of the salt (and of the acid) at any point is calculated from the volume of alkali added due to allowance being made for the total volume of the solution. The initial pH of 0.1M acetic acid is computed from equation (6); the dissociation of the acid is relatively so small that it may be neglected in expressing the concentration of acetic acid. The pH values at other points on the titration curve are similarly calculated. After the equivalence point has been passed, the solution contains an excess of OH⁻ ions which will repress the hydrolysis of the salt; the pH may be assumed, with sufficient accuracy due to the excess of base present, so that in this region, the titration curve will almost coincide with that 0.1 M hydrochloric acid (Fig. **2**).

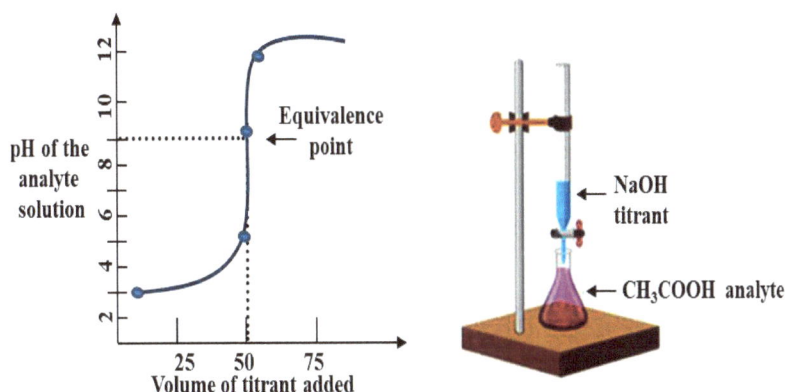

**Fig. (2).** Titration of weak acid with strong base along with graph.

## Titration of a Strong Acid with a Weak Base

The titration of 100 mL of 0.1M aqueous ammonia (Kb= 1.85 x with 0.1M) hydrochloric acid at the ordinary laboratory temperature. For other concentrations, the pH may be calculated with sufficient accuracy as follows:

$$\text{Or } [OH\text{-}] = [NH,] \times K, /[NH:]$$

$$\text{Or pOH} = \log [Salt]/ [Base] + pK$$

$$\text{Or pH} = pKw - pK, - \log [Salt]/ [Base]$$

After the equivalence point has been reached, the solution contains an excess of $H^+$ ions, hydrolysis of the salt is suppressed, and the subsequent pH changes may be assumed, with sufficient accuracy, to be those due to the excess of acid present. The results computed in the above manner are represented graphically the results for the titration of 100 mL of a 0.1 M solution of a weaker base. It is clear that neither thymolphthalein nor phenolphthalein can be employed in the titration of 0.1 M aqueous ammonia (Fig. **3**).

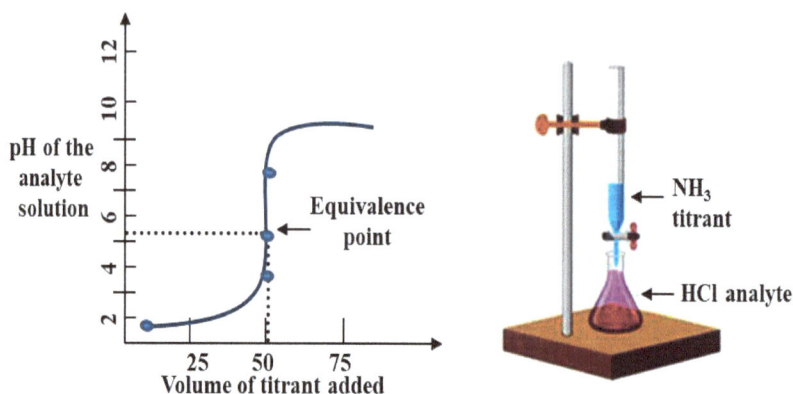

**Fig. (3).** Titration of strong acid with a weak base along with graph.

The equivalence point is at pH 5.3, and it is necessary to use an indicator with a pH range on the slightly acid side (3-6.5), such as methyl orange, methyl red, bromophenol blue, or bromocresol green. The last-named indicators may be utilised for the titration of all weak bases.

## Titration of a Weak Base with a Weak Acid

The neutralization curve up to the equivalence point is almost identical with that using 0.1 M sodium hydroxide as the base; beyond this point, the titration is virtually the addition of 0.1 M aqueous ammonia solution to 0.1 M ammonium acetate solution and equation (1:1) is applicable to the calculation of the pH. The titration curve for the neutralisation of 100 mL of 0.1 M acetic acid with 0.1 M aqueous ammonia at the laboratory temperature.

The chief feature of the curve is that the change of pH near the equivalence point and, indeed, during the whole of the neutralisation curve is very gradual. There is no sudden change in pH, and hence no sharp endpoint can be found with any simple indicator. A mixed indicator, which exhibits a sharp colour change over a very limited pH range, may sometimes be found which is suitable [21]. Thus, for acetic acid-ammonia solution titrations, neutral red-methylene blue mixed indicator may be used, but on the whole, it is best to avoid the use of indicators in titrations involving both a weak acid and a weak base (Fig. 4).

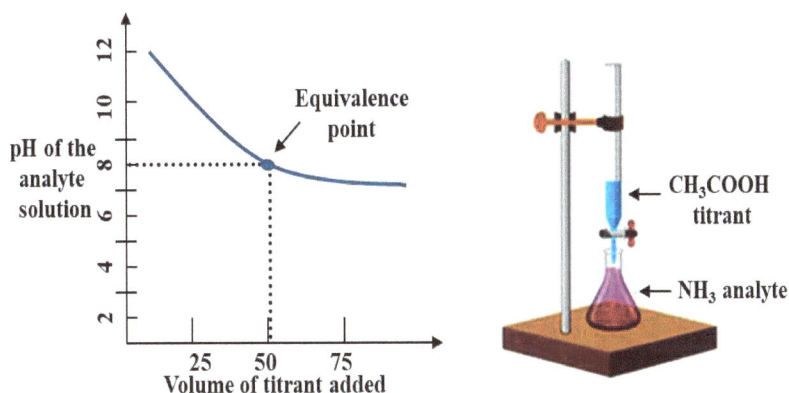

**Fig. (4).** Titration of weak acid with weak base along with graph.

This basic solution was added to the acidic solution until complete neutralization was obtained. The endpoint of titration will be detected with the help of an indicator as colour of the solution changes upon neutralization.

## CONSENT FOR PUBLICATION

Not applicable.

## CONFLICT OF INTEREST

The author declares no conflict of interest, financial or otherwise.

## ACKNOWLEDGEMENTS

Declared none.

## REFERENCES

[1]    Adams, W.K.; Wieman, C.E. Development and Validation of Instruments to Measure Learning Of Expert-Like Thinking. *Int. J. Sci. Educ.,* **2011**, *33*(9), 1289-1312.
       [http://dx.doi.org/10.1080/09500693.2010.512369]

[2]    Airasian, P.W.; Miranda, H. The Role of Assessment in the Revised Taxonomy. *Theory Pract.,* **2002**, *4*(4), 249-254.
       [http://dx.doi.org/10.1207/s15430421tip4104_8]

[3]    Anderson, L.W. *Classroom Assessment: Enhancing the Quality of Teacher Decision Making*; Lawrence Erlbaum Associates: Mahwah, NJ, **2003**.
       [http://dx.doi.org/10.4324/9781410607140]

[4]    Artdej, R.; Ratanaroutai, T.; Coll, R.K.; Thongpanchang, T. Thai Grade 11 Students' Alternative Conceptions for Acid–Base Chemistry. *Res. Sci. Technol. Educ.,* **2010**, *28*(2), 167-183.

[http://dx.doi.org/10.1080/02635141003748382]

[5]     Bell, B.; Cowie, B. The Characteristics of Formative Assessment in Science Education. *Sci. Educ.,* **2001**, *85*(5), 536-553.
        [http://dx.doi.org/10.1002/sce.1022]

[6]     Brislin, R.W. Back-Translation for Cross-Cultural Research. J. Cross-Cult. *Psy.,* **1970**, *1*(3), 185-216.

[7]     Burns, J.R. An Evaluation of 6th and 7th Form Chemistry In Terms of the Needs of the Students and the Community. In: *Report to the Department of Education*; Wellington, New Zealand, **1982**.

[8]     Chandrasegaran, A.L.; Treagust, D.F.; Mocerino, M. The Development Of A Two-Tier Multiple-Choice Diagnostic Instrument For Evaluating Secondary School Students' Ability To Describe And Explain Chemical Reactions Using Multiple Levels Of Representation. *Chem. Educ. Res. Pract.,* **2007**, *8*(3), 293-307.
        [http://dx.doi.org/10.1039/B7RP90006F]

[9]     Chandrasegaran, A.L.; Treagust, D.F.; Mocerino, M. An Evaluation of a Teaching Intervention to Promote Students' Ability to Use Multiple Levels of Representation When Describing and Explaining Chemical Reactions. *Res. Sci. Educ.,* **2008**, *38*(2), 237-248.
        [http://dx.doi.org/10.1007/s11165-007-9046-9]

[10]    Winters, R.W. Terminology of acid-base disorders. *Ann. Intern. Med.,* **1965**, *63*(5), 873-884.
        [http://dx.doi.org/10.7326/0003-4819-63-5-873] [PMID: 5848634]

[11]    Mitchell, R.A.; Carman, C.T.; Severinghaus, J.W.; Richardson, B.W.; Singer, M.M.; Shnider, S. Stability of cerebrospinal fluid pH in chronic acid-base disturbances in blood. *J. Appl. Physiol.,* **1965**, *20*(3), 443-452.
        [http://dx.doi.org/10.1152/jappl.1965.20.3.443] [PMID: 5837560]

[12]    Mitchell, R.A.; Singer, M.M. Respiration and cerebrospinal fluid pH in metabolic acidosis and alkalosis. *J. Appl. Physiol.,* **1965**, *20*(5), 905-911.
        [http://dx.doi.org/10.1152/jappl.1965.20.5.905] [PMID: 5837617]

[13]    Albert, M.S.; Winters, R.W. Acid-base equilibrium of blood in normal infants. *Pediatrics,* **1966**, *37*(5), 728-732.
        [http://dx.doi.org/10.1542/peds.37.5.728] [PMID: 5932624]

[14]    Siggaard-Andersen, O. *The Acid-Base Status of the Blood,* 2nd ed; The Williams & Wilkins Co.: Baltimore, **1964**.

[15]    Finberg, L. Pathogenesis of lesions in the nervous system in hypernatremic states. I. Clinical ovservations of infants. *Pediatrics,* **1959**, *23*(1 Pt 1), 40-45.
        [http://dx.doi.org/10.1542/peds.23.1.40] [PMID: 13613862]

[16]    Winkler, A.W.; Elkinton, J.R.; Hopper, J., Jr; Hoff, H.E. EExperimental hypertonicity: alterations in the distribution of body water, and the cause of death. *J. Clin. Invest.,* **1944**, *23*(1), 103-109.
        [http://dx.doi.org/10.1172/JCI101464] [PMID: 16695075]

[17]    Kety, S.S.; Polis, B.D.; Nadler, C.S.; Schmidt, C.F. Schmidt, C.F. The Blood Flow and the Oxygen Consumption of the Human Brain in Diabetic Acidosis and Coma. *J. Clin. Invest.,* **1948**, *27*(4), 500-510.
        [http://dx.doi.org/10.1172/JCI101997]

[18]    Steel, R.G.D.; Jorrie, J.H. *Principles and Procedures of Statistics*; Mcgraw Hill Book Co.: New York, **1960**.

[19]    Poppell, J.W.; Vaname, P.; Roberts, K.E.; Randall, H.T. The Effect of Ventilatory Insufficiency on Respiratory Compensation in Metabolic Acidosis and Alkalosis. *J. Lab.Clin.ical Med,* **1956**, *47*(6), 885-90.

[20]    M0ller, B. Hydrogen Ion Concentration in Arterial Blood. *A Clinical Study of Patients with Diabetes Mellitus and Diseases of the Kidneys, Lungs, and Heart,* Acta Med. Scand. **1959**, *165* (Suppl. 348)

[21]     Lennon, E.J.; Lemann, J., Jr Defense of hydrogen ion concentration in chronic metabolic acidosis. A new evaluation of an old approach. *Ann. Intern. Med.,* **1966**, *65*(2), 265-274.
[http://dx.doi.org/10.7326/0003-4819-65-2-265] [PMID: 5912890]

# CHAPTER 4

# Non-Aqueous Titrations

**Sunaina Aggarwal[1,*], Upasna Kaushik[1] and Shivani Sharma[1]**

*[1] RKSD College of Pharmacy, Kaithal, Haryana, India-136027*

**Abstract:** Various titrimetric methods are available for estimation of acids and bases, except very weak acids and bases which can only be estimated by either non-aqueous methods of titration or potentiometry. Non-aqueous means of titrimetric analysis involve the conversion of weak acid to a strong acid or weak base to a strong base by exerting differentiating effect when dissolved in a basic or acidic solvent. The endpoint of the titrations is determined *via* non-aqueous indicator solutions such as Crystal violet, Nile blue A, 1-Naphtholbenzein, Oracet blue B.

**Keywords:** Acidimetry, Alkalimetry, Differentiating effect, Leveling effect, Non-aqueous titration.

## INTRODUCTION

For decades, acids have been defined as substances sour in taste and turn blue litmus solution/paper red. Similarly, bases are substances bitter in taste and turn red litmus solution/paper red. Acids and bases show dissociation to give their respective ions in solution.

Compounds with high ionization are referred to as strong acids and bases while weakly or partially ionizing compounds as weak acids and bases. Several theories were forwarded to explain the chemical nature of acids and bases.

According to the Bronsted-Lowry concept, acids are proton donors, while bases are proton acceptors. Dissociation of an acid HA in a solution yields a proton $H^+$ and a conjugate base $A^-$, while a base B combine with a proton $H^+$ and forms a conjugate acid $HB^+$.

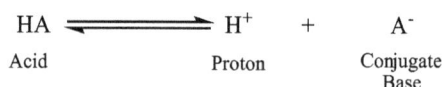

$$HA \rightleftharpoons H^+ + A^-$$

Acid        Proton      Conjugate
                         Base

---

\* **Corresponding Author Sunaina Aggarwal**: RKSD College of Pharmacy, Kaithal, Haryana, India-136027; Email: sunainaaggarwal53@gmail.com

**Anju Goyal & Harish Kumar (Eds.)**
**All rights reserved-© 2022 Bentham Science Publishers**

$$B^- \quad + \quad H^+ \quad \rightleftharpoons \quad BH$$

Base  Proton  Conjugate Acid

Similarly, self-ionization of water gives hydronium ion ($H^3O^+$) as conjugate acid and hydroxyl ion ($OH^-$) as a conjugate base.

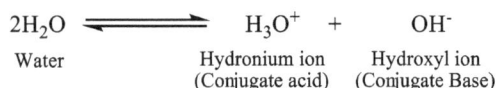

$$2H_2O \rightleftharpoons H_3O^+ \quad + \quad OH^-$$

Water  Hydronium ion (Conjugate acid)  Hydroxyl ion (Conjugate Base)

Water, being a universal solvent, can be effectively used as a medium for acid-base titration. The amphoteric nature of water, however, limits its use while titrating weak acids and bases. Considering titration of a base with an acid, both solutions being aqueous, base and hydroxyl ion ($OH^-$) will compete for the proton provided by acid. In case of a weak base, there is very low competition with solvent for proton, and hence, the base cannot be titrated effectively.

In a similar manner, when an aqueous solution of acid is being titrated by a base, it shows competition of weak acid and hydronium ion ($H_3O^+$) for the base. Thus, in an aqueous solution, weak acids or weak bases cannot be titrated effectively due to overwhelming effect of weak acid and base molecules of solvent.

This problem can be solved by replacing the solvent. If a weak base is dissolved in a non-basic solvent. The solute-solvent competition for proton can be reduced and the base can be titrated easily. Similarly, weak acid can be titrated effectively by dissolving in a non-acidic solution. Here, when water is used as a solvent it exerts a leveling effect, while when some acid or base is used as solvent, they exert differentiating effect [1 - 3, 6].

## SOLVENT SYSTEM FOR NON-AQUEOUS TITRATIONS

The acid-base property of a solvent depends solely on its tendency to accept or donate a proton. Considering the acid-base properties, H. A. Laitinen (1960) classified solvents as:

### Aprotic Solvents

Chemically neutral and virtually unreactive compounds under the conditions employed are referred to as aprotic solvents. These compounds have no tendency to accept or yield a proton. *E.g.*, carbon tetrachloride, toluene, chloroform. Aprotic solvents show low dielectric constant, do not ionize the solutes or show acid-base reaction. Such solvents are used for diluting the reaction mixtures.

## Protophilic Solvents

Substances included under protophilic solvents show a high affinity for accepting a proton *e.g.* pyridine, ammonia, acetic anhydride, ether. These solvents possess high basicity and low acidity than water.

## Protogenic Solvents

Protogenic solvents readily dissociate to yield a proton and are highly acidic for example hydrogen fluoride, sulphuric acid, and acetic acid.

## Amphiprotic Solvents

Amphiprotic solvents show the properties of both progenic and protophilic solvents, *i.e.,* they have a tendency to either accept or donate a proton for example water and alcohol [3, 4, 7].

## FACTORS AFFECTING SELECTION OF SOLVENT

Factors that are required to be considered while selecting a solvent system for non-aqueous titration includes:

## Acid-Base Character

Consider a protophillic solvent that is used to dissolve an acidic drug. Since a protophillic solvent shows poor tendency to lose a proton, while undergoing titration, it will not compete with acidic solute thus it can easily be titrated. In other words, protophillic solvents being basic in nature can be used to dissolve weakly acidic substances, thereby increasing its strength. Similarly, strength of weakly basic substances can be increased by dissolving in protogenic solvents and titrated easily without any hindrance.

## Dielectric Constant

Dielectric constant refers to the measurement of the ease by which oppositely charged ions dissociate. Mathematically, it is represented by the formula,

$$F = \frac{(q_1 q_2)}{D(r_1 r_2)^2}$$

Where F is the electrostatic force of attraction or repulsion

$q_1$ is charge on cation

$q_2$ is charge on anion

D is Dielectric constant

$r_1$ is radius of cation

$r_2$ is radius of anion

Since, electrostatic force of attraction or repulsion is inversely proportional to dielectric constant, we may say the lower the dielectric constant of a solvent, higher the electrostatic forces required to ionize it and hence less ionization and *vice-versa*.

In a non-aqueous titration, the solvent to be used must possess a lower dielectric constant, rendering it unionized, while that of solute must be high so that it can be ionized easily in solution form [5, 8 - 10].

## SOLVENTS EMPLOYED IN NON-AQUEOUS TITRATION

Both inorganic and organic solvents can be used in non-aqueous titrations. Solvents to be used must be pure, dry and of analytical grade. Most widely used solvents are discussed below:

### Glacial Acetic Acid

Glacial acetic acid is the most widely used solvent in non-aqueous titrations. Acetic acid is usually available with 0.1%and 1.0% water content, which may be reduced by adding sufficient acetic anhydride.

### Acetonitrile

Acetonitrile is widely used in conjunction with chloroform, phenol and acetic acid. It is widely used in titration of metal ethanoates with perchloric acid in order to obtain a sharp endpoint.

### Alcohols

A mixture of glycols and alcohols is used for determining the strength of organic acids and their salts. Ethylene glycol, when mixed with 2-propanol or 1-butanol is the most widely used alcoholic combination.

## Dioxane

When determining the strength of a mixture of substances, dioxane can be used, as it provides individual endpoints for each component of the mixture which is not possible with other solvents.

## Dimethylformamide

Being a protophillic solvent, dimethylformamide is used for titrating benzoic acid with amides where endpoint is difficult to obtain [3, 7].

## INDICATORS IN NON-AQUEOUS TITRATIONS

Indicators commonly used in non-aqueous titrations include:

## Crystal Violet

Used in conc. 0.5% in glacial acetic acid and shows colour change from blue in basic medium to yellowish-green in acid.

## 1-Naphtholbenzein

Used in conc. 0.2% in glacial acetic acid and shows a colour change from blue to dark green when the medium changes from basic to acidic.

## Nile Blue A

used as 1% solution in glacial acetic acid and imparts a blue colour to basic solution while blue-green colour to acidic solution.

## Oracet Blue B

Solution is prepared in glacial acetic acid, maintaining a conc. of 0.5% showing a colour change of blue to pink when transiting from basic medium to acidic medium.

## Quinaldine Red

Used in the conc 0.1% to be prepared in methanol. Quinaldine red shows a colour change from magenta in the basic medium to almost colourless in acid.

## Thymol Blue

prepared as 0.1% solution in DMF and imparts yellow colour to acidic solution while blue to base.

**Azovoilet**

used as 0.2% solution, prepared in benzene, and imparts orange colour to acid while violet to base with a pink in neutral solution [1, 4].

## ALKALIMETRY

Weakly acidic substances can be dissolved in basic or neutral solvents such as ethylenediamine, pyridine and *n*-butylamine as basic solvents increase their acidic character. Other extensively used solvents are methanol, ethanol, acetone, methyl ethyl ketone, methyl isobutyl ketone and *t*-butyl alcohol.

These substances can be analysed using potassium, sodium or lithium methoxide prepared by dissolving the metal in toluene-methanol or with tetrabutylammonium hydroxide solution in methanol. One of the drawbacks of using metal methoxides is the formation of gelatinous precipitates and their toxicity. To avoid this problem, tetrabutylammonium hydroxide was introduced as titrant.

Benzoic acid is a widely used primary standard for non-aqueous alkalimetry, while thymol blue and azovoilet are visual indicators available. In the absence of visual indicator, endpoint of the titration is determined *via* potentiometry [3, 11].

## ESTIMATION OF ACIDS

### Preparation of 0.1N Alkali Methoxide

2.3-2.6g freshly cut sodium or 0.7-1.0g of lithium using absolute methanol should be rinsed. To an ice-cold mixture of 50ml toluene and 40ml absolute methanol, the metal with continuous stirring should be added. Once the metal is dissolved, add methanol to make the solution homogenous followed by addition of toluene until the solution is cloudy. Continue addition of methanol and toluene alternately till the final volume is 1 litre. Sore the solution in pyrex or polyethylene bottle and standardize it against benzoic acid.

### Standardization of 0.1N Alkali Methoxide

Dissolve accurately weighed 200-300 mg benzoic acid in Erlenmeyer flask in 25ml *N,N*-dimethyl formamide (DMF). Add a few drops of thymol blue, pass a stream of nitrogen and titrate using alkali methoxide solution until colour of the solution changes to blue. Take minimum three concordant readings and a blank reading using 25 ml DMF. The chemical reactions involved are shown below:

$$2Na \ + \ 2CH_3OH \longrightarrow \ 2CH_3ONa \ + \ H_2$$

$$C_6H_5COOH \ + \ H{-}CON(CH_3)_2 \longrightarrow \ HCON^+H(CH_3)_2 \ + \ C_6H_5COO^-$$

$$CH_3ONa \rightleftharpoons \ CH_3O^- \ + \ Na^+$$

$$HCON^+H(CH_3)_2 \ + \ CH_3O^- \longrightarrow \ HCON(CH_3)_2 \ + \ CH_3OH$$

Net
Reaction:    $C_6H_5COOH \ + \ CH_3ONa \longrightarrow \ C_6H_5COONa \ + \ CH_3OH$

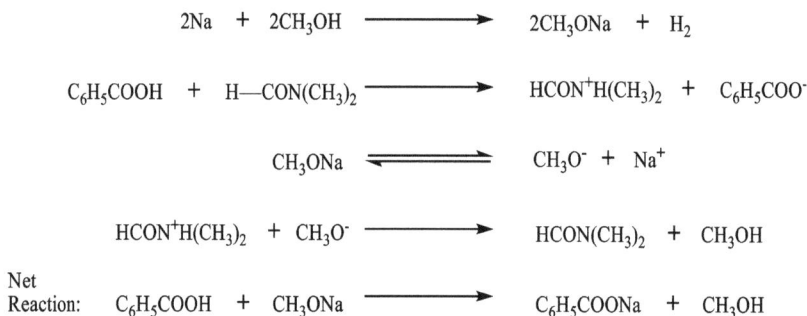

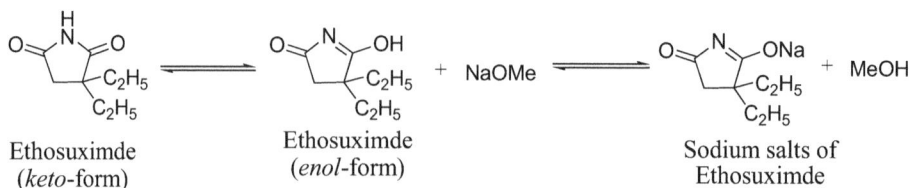

For calculations,

1 ml 0.1N lithium methoxide = 0.01221 g of benzoic acid

## ASSAY OF ETHOSUXIMIDE

**Materials Required:** Ethosuximide: 0.2 g; dimethylformamide: 50 ml; azo-violet (0.1% w/v in DMF): 2 drops; sodium methoxide 0.1 N.

**Procedure:** Weigh accurately about 0.2 g, dissolve in 50 ml of dimethylformamide, add 2 drops of azo-violet solution and titrate with 0.1 N sodium methoxide to a deep blue endpoint, taking precautions to prevent absorption of atmospheric carbon dioxide. Perform a blank determination and make any necessary corrections. Each ml of 0.1 N sodium methoxide is equivalent to 0.01412 g of $C_7H_{11}NO_2$.

Equations:

Ethosuximde      Ethosuximde      Sodium salts of
(*keto*-form)      (*enol*-form)      Ethosuximde

Calculations:

Therefore,

141.17 g $C_7H_{11}NO_2 \equiv$ NaOMe $\equiv$ H $\equiv$ 1000 ml N

or

$0.01417$ g $C_7H_{11}NO_2 \equiv 1$ ml $0.1$ N NaOMe

## Preparation of 0.1N Tetrabutylammonium Hydroxide

Dissolve 80gms of accurately weighed tetrabutylammonium iodide ($Bu_4NI$) in 180ml of analytical reagent grade anhydrous methanol taken in a stoppered flask. Place the flask in an ice bath, add 40gms of powdered silver oxide and shake occasionally for one hour. Filter the solution through sintered glass crucible. Wash the flask and residue thrice with 50ml of cold and dry benzene. Combine the filtrate and washings and dilute to 2 litres with dry benzene. The chemical reaction involved is shown below.

$$2Bu_4NI + Ag_2O + H_2O \longrightarrow 2Bu_4NOH + 2AgI$$

## Standardization 0.1N Tetrabutylammonium Hydroxide

Weigh accurately 0.1gm of benzoic acid previously dried at 100°C for 30 min and dissolve in 50ml of analytical reagent grade pyridine. Insert the electrodes in the solution, start the stirrer and adjust the millivolt scale reading to zero.

Titrate potentiometrically with tetrabutylammonium hydroxide solution. Also prepare millivolt-scale versus volume of titrant added graph [11].

For calculations,

Therefore, $C_6H_5COOH \equiv H \equiv 1000$ ml N

or $0.01221$ g $C_7H_6O_2 \equiv 1$ ml of $0.1$ N $Bu_4NI$

## ASSAY OF CHLORTHALIDONE

**Materials Required:** Chlorthalidone: 0.3 g; pyridine (dehydrated): 50 ml; 0.1 N tetrabutylammonium hydroxide.

**Procedure:** Weigh accurately about 0.3 g and dissolve in 50 ml of dehydrated pyridine. Titrate with 0.1 N tetrabutylammonium hydroxide, determining the end point potentiometrically and protecting the solution and titrant from atmospheric carbon dioxide throughout the determination. Perform a blank determination and make any necessary correction.

Each ml of 0.1 N tetrabutylammonium hydroxide is equivalent to 0.03388 g of $C_{14}H_{11}ClN_2O_4S$.

Equations:

Calculations:

Therefore, $C_{14}H_{11}ClN_2O_4S \equiv Bu_4N^+OH^- \equiv H \equiv 1000$ ml N

or 338.76 g $C_{14}H_{11}ClN_2O_4S \equiv 1000$ ml N

or 0.0338 g $C_{14}H_{11}ClN_2O_4S \equiv 1$ ml 0.1 N

## ACIDIMETRY

In a manner similar to alkalimetry, neutral or acidic solvents are used for salvation of weakly basic substances. Glacial acetic acid is the solvent of choice in non-aqueous acidimetry.

However, in case of very weak bases such as amides, acetic anhydride is used. Other solvents employed in non-aqueous titration are dioxane, perchloric acid, acetonitrile, benzene, chloroform, a mixture of glycol-hydrocarbon (*e.g.* 1:1 solution of propylene glycol and chloroform or that of ethylene glycol and *n*-butanol).

Perchloric acid, being the strongest acid available, is the most commonly used titrant in non-aqueous acidimetry. Commercially available perchloric acid is a concentrated aqueous solution. Water is removed from the solution by adding acetic anhydride, in order to increase its sensitivity towards titrations. Primary standard used in non-aqueous acidimetry is potassium biphthalate. The chemical reaction between potassium biphthalate and perchloric acid is shown below:

Other substances used as primary standard are tris(hydroxymethyl) aminomethane, sodium carbonate and diphenylguanidine. Crystal violet, methyl violet, *p*-naphtholbenzein, quinaldine red, malachite green are the visual indicators frequently used in the non-aqueous titration of weak bases. In the absence of visual indicator, endpoint of the titration is determined *via* potentiometry [1, 3].

# ESTIMATION OF AMINES AND AMINE SALTS OF ORGANIC ACIDS

## Preparation of 0.1N Perchloric Acid

Take 1000ml volumetric flask containing 900ml of glacial acetic acid and slowly add 11ml of 60% perchloric acid to it with continuous stirring. Then add 36.0ml of reagent grade acetic anhydride, mix and dilute to 1000ml with acetic acid. Allow the solution to stand at room temperature for 24hrs before standardization.

In case of availability of 72% perchloric acid, the solution can be prepared by slowly adding 8.5ml of 72% perchloric acid to 900ml of glacial acetic acid and 21ml of acetic anhydride with continuous stirring, further diluting to 1000ml.

## Standardization of 0.1N Perchloric Acid

Potassium hydrogen phthalate may be used as a primary standard against acetous perchloric acid. Accurately weigh about 0.5gm Potassium hydrogen phthalate in a dry Erlenmeyer flask and dissolve it in 25ml of glacial acetic acid with mild heating. Cool the solution to room temperature and add 2-3 drops of either 0.5% w/v acetous crystal violet or 0.5% w/v acetous oracet blue. Attach a reflux condenser fitted with silica gel drying tube and warm until the salt has been completely dissolved. Bring the solution to room temperature and titrate with 0.1N perchloric acid. Record the titrant temperature. Take minimum three concordant readings and a blank reading [3].

For calculations,

1ml 0.1N Perchloric acid = 0.02041g of Potassium hydrogen phthalate

## ASSAY OF METHYLDOPA

Reaction between methyldopa and perchloric acid may be expressed as:

**Procedure:** Dissolve accurately weighed 0.4gms methyldopa in 15ml of anhydrous formic acid, 30 ml of glacial acetic acid, and 30ml of dioxane. Titrate the solution against perchloric acid using acetous crystal violet as an indicator. Perform a blank titration and make necessary corrections. The chemical reaction involved is shown below [5].

For calculations,

1ml 0.1N Perchloric acid = 0.02112g of Methyl dopa

## ASSAY OF NITRAZEPAM

Reaction between nitrazepam and perchloric acid may be expressed as:

**Procedure:** Dissolve accurately weighed 0.5gm of nitrazepam in 50ml acetic anhydride and titrate with 0.1M perchloric acid using nile blue as an indicator. Perform a blank titration and make necessary corrections. The chemical reaction involved is shown below [1].

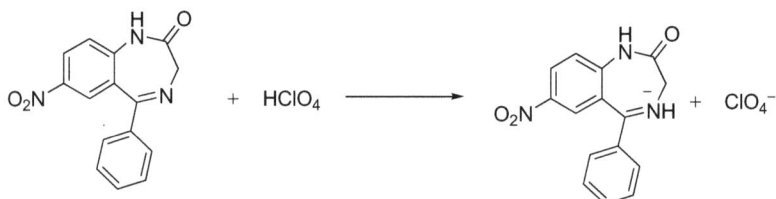

For calculations,

1ml 0.1M Perchloric Acid = 0.0281g of Nitrazepam

## ESTIMATION OF HALOGEN ACID SALTS OF BASES

### Assay of Amitriptyline Hydrochloride

General reaction between Amitriptyline Hydrochloride and perchloric acid may be expressed as:

**Procedure:** Dissolve in 50ml of acetic acid accurately weighed 1.0gm of Amitriptyline hydrochloride. Warm the solution to aid in dissolution if required. Cool to room temperature and add 10ml mercuric acetate solution, 0.1ml crystal violet solution and titrate with 0.1N perchloric acid until end point. Perform a blank titration and make necessary corrections. The chemical reaction involved is shown below [5].

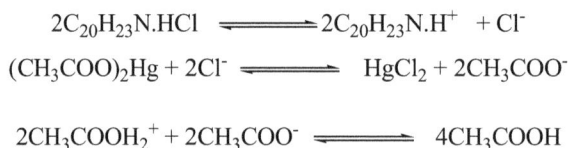

$$2C_{20}H_{23}N.HCl \rightleftharpoons 2C_{20}H_{23}N.H^+ + Cl^-$$

$$(CH_3COO)_2Hg + 2Cl^- \rightleftharpoons HgCl_2 + 2CH_3COO^-$$

$$2CH_3COOH_2^+ + 2CH_3COO^- \rightleftharpoons 4CH_3COOH$$

For calculations,

1ml 0.1N Perchloric acid = 0.03139g of Amitriptyline hydrochloride

## ADVANTAGES OF NON-AQUEOUS TITRATIONS

Various advantages of non-aqueous titrations are listed below [9, 10, 12]:

- Dissolution of weakly soluble drugs can be improved.
- Water-insoluble compounds can be easily analyzed.
- Weak acids and bases can be titrated using either visual indicators or electrochemical methods.

## LIMITATIONS OF NON-AQUEOUS TITRATIONS

The factors which limit the use of non-aqueous titrations as an analytical method involves [9, 10, 12]:

- Maintenance of constant temperature of titrant and titrate throughout the titration.
- Protection of titrant, titrate and indicator from carbon dioxide and moisture.
- Environmental pollution by volatile solvents.

## CONCLUSION

Non-aqueous means of titrimetric analysis involve conversion of weak acid to a strong acid or weak base to a strong base by exerting differentiating effect when dissolved in a basic or acidic solvent. End point of the titrations is determined *via* non-aqueous indicator solutions such as Crystal violet, Nile blue A, 1-Naphtholbenzein, and Oracet blue B. Aprotic solvent, Protophillic solvent, protogenic solvents and amphiprotic solvents have been used for titrimetric purpose. Mostly these methods have been employed for the estimation of acids, assay of ethosuximide, assay of chlorthalidone, assay of nitrazepam, estimation of amines and amines salt of organic acids, *etc*. This method has also been associated with some limitations like maintenance of constant temperature, protection of titrant, titrate and indicator from moisture and environmental pollution.

## CONSENT FOR PUBLICATION

Not applicable.

## CONFLICT OF INTEREST

The author declares no conflict of interest, financial or otherwise.

## ACKNOWLEDGEMENTS

Declared none.

## REFERENCES

[1]     Beckett, A.H.; Stenlake, J.B. *Practical Pharmaceutical Chemistry-Part 1,* 4th ed; CBS Publishers and Distributors, **2003**, pp. 165-175.

[2]     Kamboj, P.C. *Pharmaceutical Analysis Theory and Practicals,* 2nd ed; Vallabh Publications, **2007**, Vol. 1, pp. 308-336.

[3]     Connors, K.A. *A Textbook of Pharmaceutical Analysis, 3rd ed.; reprint, Wiley India Pvt. Ltd*; , **2009**, pp. 46-65.

[4]     Vogel, A.I. *Elementary Practical Organic Chemistry: Part 3: Quantitative Organic Analysis, 2nd edition, reprint, CBS Publishers and Distributors*; , **2002**, pp. 663-675.

[5]     Kar, A. *Pharmaceutical Drug Analysis, 2nd ed.; reprint, New Age International (P) Limited, Publishers*; , **2007**, pp. 106-121.

[6]     Chatwal, G.R. *Pharmaceutical Chemistry- Inorganic,* 4th ed; Himalaya Publishing House, Pvt. Ltd., **2008**, Vol. 1, pp. 74-78.

[7]     Jeffery, G.H.; Bassett, J.; Mendham, J.; Denney, R.C. *Vogel's Textbook of Quantitative Chemical Analysis,* 5th ed; Longman Scientific and Technical, **1989**, pp. 281-284.

[8]     Verma, R.M. Analytical Chemistry- Theory and Practice 3rd ed. reprint, CBS Publishers and Distributors, **2007**, 195-202.

[9]     Sagar, G.V. *Basics of Drug Analysis*; Pharma Med. Press, **2009**, pp. 157-166.

[10]    Siddiqui, A.A.; Siddiqui, S. *Pharmaceutical Analysis,* 1st ed; CBS Publishers and Distributors, **2006**, pp. 57-70.

[11]    Garratt, D.C. *The Quantitative Analysis of Drugs,* 3rd ed; CBS Publishers and Distributors, **2005**, pp. 792-795.

[12]    Kasture, A.V.; Wadodkar, S.G.; Mahadik, K.R.; More, H.N. *Pharmaceutical Analysis,* 7th ed; Nirali Prakashan, **2003**, Vol. 1, pp. 73-84.

# CHAPTER 5

# Redox Titrations

**Sapna Kumari[1,*], Anju Goyal[1]** and **Madhukar Garg[1]**

[1] *Chitkara College of Pharmacy, Chitkara University, Rajpura, Punjab, India*

**Abstract:** Redox titrations are the titrimetric method developed to know the concentration of the analyte by creating redox reaction among titrant and analyte. The basic principle involved in the redox titration is the oxidation-reduction reaction, in which electron transfer from one reactant to another reactant takes place. Oxidation means loss of electron and reduction means gain of electron. These must happen at the same time, when a substance loses electrons, there must be some other substance to accept those electrons. The main applications associated with redox titrations are determining the reduction potential of sHdrA flavin from *Hyphomicrobium denitrificans,* water in non-aqueous solutions, dissolved oxygen in water and determination of alcohol content in the whiskey bottles, *etc.*

**Keywords:** Oxidation, Redox titrations, Reduction, Titrimetric methods.

## INTRODUCTION

Redox titrations are the titrimetric method based upon the electron transfer between the reactants or change in the oxidation number of the reactant. Popularly, these are well known as oxidation-reduction titrations, as these involve the titration of the reducing agent with a standard solution of an oxidizing agent and *vice versa*. Most commonly, this method is used in the laboratory for determining the concentration of a given analyte *via* triggering a redox reaction between the analyte and the titrant. Sometimes, these types of titrations needed a potentiometer or a redox indicator to identify unknown analytes [1].

## HISTORY AND DEVELOPMENT OF REDOX TITRATION METHOD

Redox titrations were announced just after the progression of acid–base titrimetry. The initial methods acquired the benefit of the oxidizing power of chlorine. In 1787, Claude Berthollet introduced a new method for the analysis of chlorine water (a mixture of $Cl_2$, HOCl, and HCl) quantitatively based on its capability to oxidize the indigo dye solution, which is colourless in its oxidized state. In 1814,

* **Corresponding author Sapna Kumari:** Chitkara College of Pharmacy, Chitkara University, Rajpura, Punjab, India; E-mail: ms.sapnakumari92@gmail.com

**Anju Goyal & Harish Kumar (Eds.)**

**All rights reserved-© 2022 Bentham Science Publishers**

Joseph Louis Gaylussac established a similar new method for chlorine in bleaching powder and endpoint was noticed visually in both methods. Because of the oxidation of indigo, the solution remained clear before the equivalence point, whereas solution got a permanent colour, because of unreacted indigo after the equivalence point.

Moreover, with the introduction of oxidizing titrant and reducing titrant such as $MnO_4^-$, $I_2$ and $Cr_2O_7^{2-}$ as and $S_2O_3^{2-}$ and $Fe^{2+}$ respectively, the demand and numbers of redox titrations increased enormously in the mid-1800s. Apart from the successful development of the new titrant, some limitations still exist just because of the lack of suitable indicators required for the titrations. The titrants having different coloured oxidized and reduced forms act as a self-indicator in the redox titration. For instance, the purple coloured $MnO_4^-$ ion acts as self-indicator, while its reduced form $Mn^{2+}$ is colourless. Thus, to improve this method, the development of a new visual indicator was mandatory at that time. The first indicator diphenylamine was developed in 1920s, followed by the development of other such indicators making the redox titration a progressive method [1, 2].

## Principle and Theory

The basic principle involved in redox titration is the oxidation-reduction reactions, in which electron transfer from one reactant to another reactant takes place. Oxidation involved loss of electron and reduction involved gain of electron. These must happen at the same time; when a substance loses electrons, then there must be some other substance to accept those electrons (Fig. **1**).

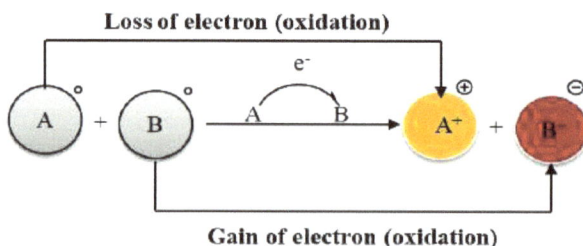

**Fig. (1).** Showing mechanism of redox titration.

In the diagram provided above, it is illustrated that an electron was removed from reactant A, and the reactant got oxidized. Likewise, reactant B was handed an electron and thus got reduced. The process of loss of electrons and increase in the oxidation state of a given reactant is called oxidation whereas, gain of electrons and decrease in the oxidation state of a reactant is called reduction. Oxidation state or oxidation number is the total number of electrons either loose or gain by an atom in order to make a chemical bond with another atom. For example, $Fe^{3+}$ has oxidation number 3, which means it has 3 electrons to make a chemical bond. In the redox reactions, the compounds that accept electrons and endure reduction are called oxidizing agents. The electron donating species that surrender electrons, termed as reducing agents, go through oxidation. From these details, it can be demonstrated that redox reactions can be fragmented into two half, reduction half reaction and the oxidation half reaction.

Examples: The assay of $FeSO_4$ using ceric sulphate. The equation involved in the reaction is:

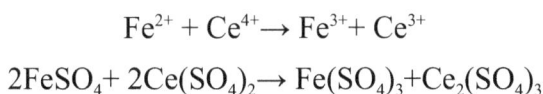

$$Fe^{2+} + Ce^{4+} \rightarrow Fe^{3+} + Ce^{3+}$$

$$2FeSO_4 + 2Ce(SO_4)_2 \rightarrow Fe(SO_4)_3 + Ce_2(SO_4)_3$$

This equation can be fragmented into two half, one half represented by ferrous ion which lose electrons and the other half shown by cerric ion which gains electrons. Oxidation means loss of electron, removal of hydrogen and addition of oxygen.

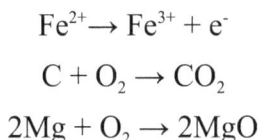

$$Fe^{2+} \rightarrow Fe^{3+} + e^-$$

$$C + O_2 \rightarrow CO_2$$

$$2Mg + O_2 \rightarrow 2MgO$$

While reduction removal of oxygen and addition of hydrogen;

$$CuO + H_2 \rightarrow Cu + H_2O$$

A redox titration is one in which the substance to be determined is either oxidized or reduced by means of the solution with which the titration is made [2].

## OXIDIZING AND REDUCING AGENTS

### Oxidizing Agents

Potassium permanganate, Potassium dichromate, Potassium bromate, Potassium iodate, Cerric sulphate, bromate-bromide mixture, Iodine, Potassium Periodate, Lead dioxide, *etc.*

## *KMnO₄ (Potassium Permanganate)*

$KMnO_4$ is the most imperative titrant, as it acts as a self-indicator in the redox titrations. It is a strong oxidizing agent and a secondary standard and its solutions can be standardized against sodium oxalate, arsenious oxide, oxalic acid, pure Mohr's salt, *etc*. Potassium permanganate acts as an oxidizing agent in all media like acidic, basic and neutral, but acts as a strong oxidizing agent in the acidic media.

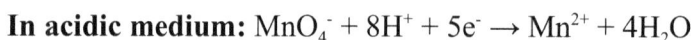

**In acidic medium:** $MnO_4^- + 8H^+ + 5e^- \rightarrow Mn^{2+} + 4H_2O$

Equivalent wt. of $KMnO_4$ = Molecular wt/5

Equivalent weight = 31.6.

It can oxidize: oxalate, $Fe^{2+}$, ferrocyanide, $H_2O_2$, $As^{3+}$ and $NO_2^-$.

**In alkaline medium:** $MnO_4^- + e^- \rightarrow MnO_4^{2-}$

Permanganate ion manganate ion

Equivalent wt. of $KMnO_4$ = Molecular wt/1

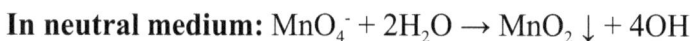

**In neutral medium:** $MnO_4^- + 2H_2O \rightarrow MnO_2 \downarrow + 4OH$

Equivalent wt. of $KMnO_4$ = Molecular wt/3

## *K₂Cr₂O₇ (Potassium Dichromate)*

It is a highly pure and stable primary standard solution.

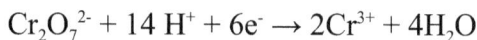

$$Cr_2O_7^{2-} + 14 H^+ + 6e^- \rightarrow 2Cr^{3+} + 4H_2O$$

## *Iodine Solution*

Iodine is very less soluble in water and its aqueous solution has appropriate vapour pressure. It can be used as standardized against standard $Na_2S_2O_3$.

$$I_2 + 2e^- \rightarrow 2I$$

## *Potassium Iodate*

It is a highly pure strong oxidizing agent.

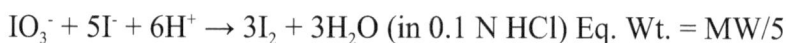

$$IO_3^- + 5I^- + 6H^+ \rightarrow 3I_2 + 3H_2O \text{ (in 0.1 N HCl) Eq. Wt.} = MW/5$$

$$IO_3^- + 2I_2 + 6H^+ \rightarrow 5I^+ + 3H_2O \text{ (in 4-6 N HCl) Eq. Wt.} = MW/4$$

$$IO_3^- + 2I + 6H^+ \rightarrow 3I^+ + 3H_2O \text{ Eq. Wt.} = MW/4$$

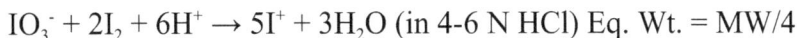

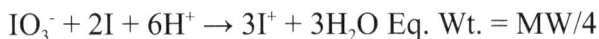

### *Bromate-bromide Mixture*

Upon acidification of this mixture, bromine is released.

$$BrO_3^- + 5\ Br^- + 6H^+ \rightarrow 3Br_2 + 3H_2O$$

It is used for the determination of phenols and primary aromatic amines:

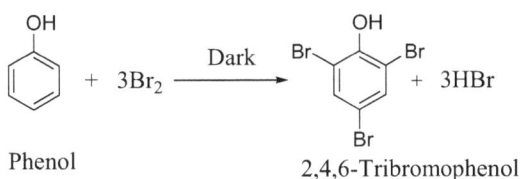

Phenol     2,4,6-Tribromophenol

The excess bromine is determined through:

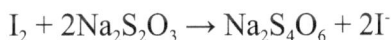

$$Br_2 + 2I^- \rightarrow I_2 + 2Br^-$$

$$I_2 + 2Na_2S_2O_3 \rightarrow Na_2S_4O_6 + 2I^-$$

## REDUCING AGENTS

Ferrous sulphate, Stannous chloride, Mohr's salt Sodium thiosulphate, Oxalic acid, Sodium arsenite, Titanous sulphate, *etc.*

## Mohr's Salt (FeSO$_4$·(NH$_4$)$_2$SO$_4$·6H$_2$O

$$Fe^{2+} \rightarrow Fe^{3+} + e^-$$

## Oxalic Acid (H$_2$C$_2$O$_4$·2H$_2$O)

$$Cr_2O_7^{2-} \rightarrow 2CO_2 + 2e^-$$

## Sodium Thiosulphate (Na$_2$S$_2$O$_3$·5H$_2$O)

$$2S_2O_3^{2-} \rightarrow S_4O_6^{2-} + 2e^-$$

# EQUIVALENT WEIGHT OF OXIDIZING AND REDUCING AGENTS

Equivalent weight of oxidizing and reducing agents can be calculated by the following formula.

$$\text{Equivalent weight} = \frac{\text{molecular weight}}{\text{electrons charge}}$$

The equivalent weight of the reducing agents can be elaborated as the weight that loses electrons equivalent to 96500 C whereas, the equivalent weight of the oxidizing agent is the weight that gains electrons equivalent to 1 faraday.

Example:

$$Fe^{2+} \rightarrow Fe^{3+} + e^{-}$$

In this equation, the equivalent weight of ferrous is 151.919.

$$Ce^{4+} + e^{-} \rightarrow Ce^{3+}$$

In the above equation, the equivalent weight of $CeSO_4$ is 332.24.

## OXIDATION NUMBER (O.N.)

In simple terms, oxidation number is the number that is allowed to the element in a chemical reaction. Basically, it is the number of electrons that an atom shares in a molecule either by losing or gaining electrons to form a chemical bond in a chemical reaction. It is also called an oxidation state. Major conditions required for assigning oxidation state to the atoms:

1. Atom should bond to heteroatom

2. Atom should always form an ionic bond, either by gaining or loosing electron, regardless of the actual nature of bonding.

Let us discuss an example of tetraoxoplatinum $(PtO_4)^{2+}$, a complex cation, in which oxidation state of platinum is 10, which is a maximum oxidation state shown by any atom. It means it can lose 10 electrons for ion formation, which requires much higher ionization energy. So, the removal of 10 electrons is only hypothetical, not practical. Practically, oxidations state larger than 3 is impossible.

Oxidation state or oxidation number of an ion or atom can be assigned by:

1. Adding the constant oxidation states of the atoms or ions in molecule/ion that are bonded to it.

2. Equating the total oxidation state of the ions/molecules to the total charge of the ion and molecule.

Oxidation number can be calculated as:

Example: O.N. of potassium permanganate ($KMnO_4$) = Sum of oxidation number of (K + Mn + 4O) = 0 and Oxidation number of permanganate ion $(MnO_4)^-$ = Sum of oxidation number of (Mn + 4O) = -1

## Rules of Assigning Oxidation Number

i. The oxidation number of an element in its elementary form is zero. For example, $H_2$, $O_2$, $N_2$, *etc.* have oxidation number equal to zero.
ii. In the compounds, oxygen has oxidation number -2. Though, some exceptions exist there. For example, compounds such as peroxides $Na_2O_2$, $H_2O_2$.

In $OF_2$ Oxidation number of oxygen = $-1$

In $O_2F_2$ O.N. of oxygen = +2

O.N. of oxygen = +1

i. In non-metallic compounds of hydrogen like HCl, $H_2S$, $H_2O$ oxidation number of hydrogen = + 1 but in metal hydrides oxidation number of hydrogen = -1[LiH, NaH, $CaH_2$, *etc.*]
ii. In compounds of metals and non-metals metals have positive oxidation number while non-metals have negative oxidation number. For example, In NaCl. Na has +1 oxidation number while chlorine has -1.
iii. The oxidation state/number of neutral atoms and molecules is zero, as they have zero net charge. For example, oxidation states atoms and molecules like magnesium, iron, sodium, oxygen, water, ammonia, chlorine, potassium permanganate, methane are zero.
iv. Oxidation number/state of atom in homopolar molecule is zero.
v. The oxidation state of charged ion = net charge of the ion. Table **1** illustrated some compounds and oxidation number of atoms and molecules.

**Table 1. List of oxidation number of some substances.**

| Substance | Oxidation Number |
|---|---|
| Alkali earth metal (Li, Na, K, Rb, Fr, Cs) | +1 |

*(Table 1) cont.....*

| Substance | Oxidation Number |
|---|---|
| Alkaline earth metal (Be, Mg, Ca, Sr) | +2 |
| Hydrogen in proton | +1 |
| Boron family (B, Al, Ga, In, Tl) | +3 |
| Oxygen in oxide ion ($O^{2-}$) | -2 |
| Oxygen in peroxide ion ($O-O^{2-}$) | -1 |
| NaCl | Na =+1, Cl = -1 |
| $NH_3$ | N = -3, H = +1 |
| $H_2O_2$ | H = +1, O = -1 |
| LiH | Li = +1, H= -1 |
| $K_2CrO_4$ | K = +1, Cr = +6, O = -2 |
| $KClO_3$ | K = +1, Cl = +5, O = -2 |
| $SO_4^{2-}$ | O = -2, S = +6 |

## PRIMARY STANDARD

Potassium iodate, iodine, potassium bromates, arsenious oxide, sodium oxalate, and potassium dichromate are an example of some primary standards used for analysis purpose.

### Preparation of Primary Standard

Mostly, the direct method is used in which dried substances are dissolved into the water in a known concentration and a volume is made in the measuring flask in a concentration term as molarity such as 0.01, 0.1M. Equivalent is not a static measure because if a solution is 0.1 N in a redox reaction, it may not have the same normality in the same other reaction.

## FACTORS AFFECTING REDOX TITRATIONS

The most prominent factor affecting redox titration is pH. For example: In acidic media potassium permanganate is a strong oxidizing agent, but in neutral and alkaline media, it is a weak oxidant. Likewise, in strong acid, potassium permanganate fades soon, while in weak acid, it fades slowly.

### Redox Indicators

Redox indicators, also termed oxidation-reduction indicators, are used in laboratories to track the progression of redox reactions and to indicate the endpoint of redox titrations. It should also produce a rapid change in the electrode potential in the proximity of the equivalence point during a redox titration, which is possible only if the indicator itself is capable of undergoing oxidation and reduction process and both the forms should have a distinct colour. The colour change occurred within a specific redox potential transition range. If the reducing volumetric solution is used, then the redox potential of the indicator must be lower than the potential of the solution, while, in case of oxidizing volumetric solution, the redox potential must be lowered [3, 4].

$$In_{oxd} + ne = In_{red}$$

At potential E the ratio of the concentration of two forms is given by the Nernst equation

$$E_T = E^\circ + RT/nF \ln[In_{oxd}]/[I_{red}]$$

## *Types of Redox Indicators*

### *Based on the Addition of Indicator*

**i. Self-indicator:** Many times, titrant itself is very intense in colour, so that even a single minute drop of it can develop a significant colour to the reaction mixture. For instance, we can consider potassium permagnate permanganate here, which is a self-indicator. Usually, self-indicators possessed a very strong colour, because of charge transfer transition occurred in them. Example included, $KMnO_4$-endpoint is pink to colourless, Iodine- end point is brown to black *etc.*

**ii. Internal Indicator:** Such indicators are added to the reaction mixture and should have lower reduction potential values than the analyte system so that it can react only with the titrant during reaction when all the analyte has been consumed. These indicators developed a noticeable colour change in the reaction mixture, indicating the progress in the reaction— for example, Phenanthroline blue, Methylene blue, *etc.*

**iii. External Indicators:** These are the indicators which are added to the reaction mixture externally on a grooved tile, not added to the reaction mixture on the whole. These indicators are used when a suitable indicator is not available for a reaction. The completion of the reaction was indicated by physical and chemical interactions of the indicator with the analyte, not through the redox reactions. The reaction between the analyte and the indicator sometimes may be irreversible and sometimes may result in precipitation. For example, Ferrous ions in dichromate solution show the Prussian blue colour with potassium ferricyanide solution.

## *Based on the Nature of the Indicator*

**i. Metal organic complexes**: Example: Phenanthroline shows the colour change from blue to red.

**ii. Free organic complexes**: Example: Methylene blue shows the colour change from blue to colourless.

## *Based on the Dependence on pH*

**i. pH independent indicators:** 2,2'-bipyridine shows the colour change from colourless to yellow, 5,6-dimethylphenanthroline shows the colour change from yellow-green to red.

**ii. pH dependent indicators:** Safrannin T shows the colour change from red-violet to colourless, Neutral red shows the colour change from red to colourless.

## *Potentiometric Method (Redox Indicators)*

Redox indicators do not participate in the redox titration, but whose oxidized and reduced forms differ in colour that depends on the solution's electrochemical potential. Some examples are listed in Table **2**. This is the most precise method to estimate the progress of the reaction, as the potentiometric method deals with the measurement of the e.m.f. between the indicator electrode and the reference electrode during the redox titration. Moreover, this method tells only about the equivalence point, not the endpoint of the reaction (Fig. **2**).

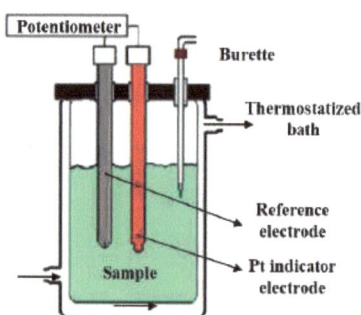

**Fig. (2).** Experimental setup for potentiometric method of endpoint determination.

**Table 2. List of Some Redox Indicators.**

| Redox Indicator | Oxidized Colour | Reduced Colour | E° (V) |
|---|---|---|---|
| indigo tetrasulfonate | blue | colourless | 0.36 |
| diphenylamine | violet | colourless | 0.75 |
| diphenylamine sulfonic acid | Red-violet | colourless | 0.85 |
| methylene blue | blue | colourless | 0.53 |
| tris(2,2'-bipyridine)iron | Pale blue | Red | 1.120 |
| Ferroin | Pale blue | Red | 1.147 |
| tris(5-nitro-1,10-phenanthroline)iron | Pale blue | Red- violet | 1.25 |

## TYPES OF REDOX TITRATIONS

## 1. Based on the Titrant Used

### *Permanganate Titration*

In this titration, the reducing substances are determined directly by the potassium permanganate and the oxidizing substances are determined indirectly.

Example:

$$2KMnO_4 + 10FeSO_4 + 8H_2SO_4 \rightarrow 5Fe_2(SO_4)_3 + K_2SO_4 + 2MnSO_4 + 8H_2O$$

### *Dichromate Titration*

potassium dichromate is a strong oxidizing agent and makes a more stable solution with any compound.

Example:

$$2K_2Cr_2O_7 + 6FeSO_4 + 7H_2SO_4 \rightarrow 3Fe_2(SO_4)_3 + K_2SO_4 + Cr_2(SO_4)_3 + 7H_2O$$

### *Iodine Titration*

### *Direct Method*

In this method, iodine is used as the titrating agent.

## *Indirect Method*

This method involved the back titration of liberated iodine with sodium thiosulphate.

## Based on the Method

### *Direct Titration*

In such types of titrations, indicator is not needed for endpoint determination as some substances are initially very coloured. For example titration of Azo dyes and quinones required no indicator to carry out reaction.

### *Back Titration*

An excess volume of the titrant solution is added to the sample solution and then the excess titrant is back titrated with the other titrant solution. Example: Chloramphenicol is titrated by this method.

## REDOX TITRATION CURVE

To assess redox titration, one must know about the shape of its titration curve. In the redox titrations, the electrochemical potential has been monitored conveniently to know the progress of reaction. From Nernst equation, we can relate the relationship between the electrochemical potential and the concentration of the reactant and products that contribute to the chemical reaction.

Let us consider an example in which an analyte in the reduced form ($A_{red}$) has been titrated with the titrant in the oxidized form ($T_{ox}$). The redox equation is

$$A_{red} + T_{OX} \rightarrow T_{red} + A_{OX}$$

The electrochemical potential for the reaction can be calculated by the difference between the reduction potential for the reduction and oxidation half reactions, so

$$E_{rxn} = E_{TOX/Tred} - E_{AOX/Ared}$$

After addition of titrant to the reaction mixture, led to the establishment of an equilibrium state between the analyte and the titrant. Therefore, at equilibrium, the electrochemical potential of the reaction ($E_{rxn}$) is zero, and

$$E_{TOX/Tred} = E_{AOX/Ared}$$

Therefore, the electrochemical potential of either half-reaction may be used to identify the progress of the titration and completion of the reaction. Prior to the equivalence point, the titration mixture contained significant amount of reduced and oxidized form of the analyte, but very diminutive amount of unreacted titrant.

At that point, the electrochemical potential can be calculated more appropriately *via* Nernst equation for the analyte's half reaction.

$$E_{Aox/Ared} = E^o_{AoxAred} - \frac{RT}{nF} \ln \frac{A_{red}}{A_{ox}}$$

Even though, $E^o_{Aox/Ared}$ is a standard state potential for the analyte's half reaction, a matrix- dependent formal potential can be used in its place. After the equivalence point, the potential is easiest to calculate for the half-reaction of titrant using Nernst equation meanwhile, noteworthy quantities of its oxidized and reduced forms are present.

$$E_{Tox/Tred} = E^o_{ToxTred} - \frac{RT}{nF} \ln \frac{T_{red}}{T_{ox}}$$

**Calculating Titration Curve**

Let us discuss an example in which a titration curve has been calculated for the 50.0mL of 0.100M $Fe^{2+}$ with 0.100M $Ce^{4+}$ in a matrix of 1 M $HClO^{4-}$.

In this case, the reaction is,

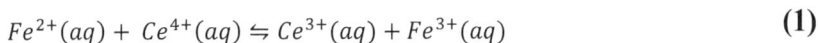

$$Fe^{2+}(aq) + Ce^{4+}(aq) \leftrightarrows Ce^{3+}(aq) + Fe^{3+}(aq) \tag{1}$$

For this reaction equilibrium constant is relatively high (appr. $6 \times 10^5$), thus we can assume that the reaction between titrant and analyte disappeared completely. After that, the volume of $Ce^{4+}$ required to achieve equivalence point was calculated and from the stoichiometry of the reaction we know

$$\text{Moles } Fe^{2+} = \text{Moles} Ce^{4+} \text{ or}$$

$$M_{Fe}V_{Fe} = M_{Ce}V_{Ce}$$

For calculating the volume of $Ce^{4+}$

$$V_{Ce} = \frac{M_{Fe}V_{Fe}}{M_{Ce}} = \frac{(0.100M)(50.0mL)}{(0.100M)} = 50.0mL$$

From the equation, we calculate equivalence point volume as 50.0mL.

From equation 1, the concentration of unreacted $Fe^{2+}$ and produced $Fe^{3+}$ before the equivalence point can be calculated. For which the potential for the analyte's half reaction should be calculated using Nernst equation.

$$E = E^o_{Fe^{3+}/Fe^{2+}} - 0.05916 \ln \frac{[Fe^{2+}]}{[Fe^{3+}]} \qquad \qquad (2)$$

The concentration of the $Fe^{2+}$ and $Fe^{3+}$ after adding 5.0 mL of titrant is

$$Fe^{2+} = \frac{moles\ unreacted\ Fe^{2+}}{Total\ Volume} = \frac{M_{Fe}V_{Fe} - M_{Ce}V_{Ce}}{V_{Ce} + V_{Ce}}$$

$$= \frac{(0.100\ M)(50.0mL) - (0.100M)(5.0mL)}{50.0mL\ +\ 5.0mL} = 8.18 \times 10^{-2}M$$

$$Fe^{3+} = \frac{moles\ Ce^{4+}\ added}{total\ volume} = \frac{M_{Ce}V_{Ce}}{V_{Fe} + V_{Ce}}$$

$$= \frac{(0.100M)(5.0mL)}{50.0mL + 5.0mL} = 9.09 \times 10^{-3}M$$

By substituting these values into equation 2, the potential for the $Fe^{3+}/Fe^{2+}$ half reaction can be calculated as follows;

$$E = +0.767\ V - 0.05916 \log \frac{8.18 \times 10^{-2}}{9.09 \times 10^{-3}} = +0.711\ V$$

At the equivalence point, moles of $Fe^{2+}$ are equal to the moles of $Ce^{4+}$ added to the reaction mixture. As the equilibrium constant for the reaction is quite large *i.e.*, 9.16, the concentration of the $Fe^{2+}$ and $Ce^{4+}$ are very small and problematic to calculate. Therefore, using Nernst equation we cannot calculate the potential at the equivalence point ($E_{eq}$). Thus, to calculate the $E_{eq}$ of the reaction, two Nernst equations of analyte's half-reaction and titrant half-reaction are combined. By doing so we found that the potentials of two half-reactions are same, so

$$E_{eq} = E^o_{Fe^{3+}/Fe^{2+}} - 0.05916 \log \frac{[Fe^{2+}]}{[Fe^{3+}]}$$

$$E_{eq} = E^o_{Ce^{4+}/Ce^{3+}} - 0.05916 \log \frac{[Ce^{3+}]}{[Ce^{4+}]}$$

Addition of these two Nernst equations resulted in;

$$2E_{eq} = E^o_{Fe^{3+}/Fe^{2+}} + E^o_{Ce^{4+}/Ce^{3+}} - 0.05916 \log \frac{[Fe^{2+}][Ce^{3+}]}{[Fe^{3+}][Ce^{4+}]} \qquad (3)$$

at the equivalence point,

$$[Fe^{2+}] = [Ce^{4+}]$$

$$[Fe^{3+}] = [Ce^{3+}]$$

Equation 3 can be simplified as, the ratio in the long term is equal to 1 and log term is zero, then;

$$E_{eq} = \frac{E^o_{Fe^{3+}/Fe^{2+}} + E^o_{Ce^{4+}/Ce^{3+}}}{2} = \frac{0.767V + 1.70V}{2} = 1.23\ V$$

After the equivalence point, the concentration of $Ce^{3+}$ and $Ce^{4+}$ is very easy to calculate. Thus, for the titrant's half-reaction, the potential can be calculated by the Nernst equation.

$$E = E^o_{Ce^{4+}/Ce^{3+}} - 0.05916 \log \frac{[Ce^{3+}]}{[Ce^{4+}]} \qquad (4)$$

Let us consider an example, that after adding 60.0mL of titrants, the concentrations of the $Ce^{3+}$ and $Ce^{4+}$ are;

$$[Ce^{3+}] = \frac{Initial\ moles\ of\ Fe^{2+}}{total\ volume} = \frac{M_{Fe}V_{Fe}}{V_{Fe} + V_{Ce}}$$

$$= \frac{(0.100M)(50.0mL)}{50.0\ mL + 60.0mL} = 4.55 \times 10^{-2}M$$

$$[Ce^{4+}] = \frac{moles\ Excess\ Ce^{4+}}{total\ volume} = \frac{M_{Ce}V_{Ce} - M_{Fe}V_{Fe}}{V_{Fe} + V_{Ce}}$$

$$= \frac{(0.100M)(60.0mL) - (0.100M)(50.0mL)}{50.0mL + 60.0mL} = 9.09 \times 10^{-3}M$$

Substituting these values in equation 4, we can calculate the potential as follows:

$$E = +1.70V - 0.05916\ log\frac{4.55 \times 10^{-2}}{9.09 \times 10^{-3}} = 1.66V$$

Furthermore, the titration curve is given below.

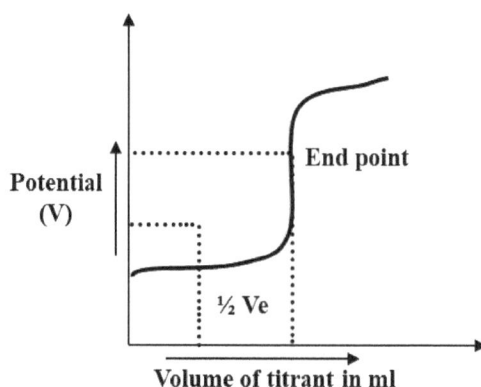

## Determining End Point

### Use of Visual Indicator

All the indicators listed above are used to determine the endpoint as their use in the titrations resulted in the colour change of the mixture. For example, $KMnO_4$, imparts different colours in the oxidized and reduced form, clearly indicating the progress of reaction. Another example is starch which forms a dark blue complex with $I^{3-}$ and becomes colourless when $I^{3-}$ got consumed. Similarly, thiocyanate, which forms a soluble red-coloured complex, $Fe(SCN)^{2+}$, with $Fe^{3+}$.

Likewise, some redox indicators that do not participate in the redox titration, but whose oxidized and reduced forms differ in colour depending on the solution's electrochemical potential. Some examples are listed in Table **2**.

### Use of Potentiometric Method

This is another popular method for locating the endpoint of a redox titration, in which an appropriate electrode is used to monitor the change in electrochemical potential when titrant is added to a solution of analyte. The endpoint can be determined through visual inspection of the titration curve. The simplest experimental design (Fig. **2**) consists of a Pt indicator electrode whose potential is governed by the analytes or titrant's redox half-reaction and a reference electrode that has a fixed potential.

## Example of Redox Titration

Let us discuss an example of titration of potassium permanganate ($KMnO_4$) against oxalic acid ($C_2H_2O_4$). The theory behind this reaction is that in the acidic media, potassium permanganate act as a strong oxidizing agent and the chemical reaction involved in this process is given below:

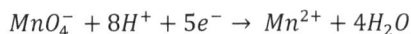

$$MnO_4^- + 8H^+ + 5e^- \rightarrow Mn^{2+} + 4H_2O$$

We also discussed previously that solution containing $MnO_4^-$ is purple in colour and becomes colourless ($Mn^{2+}$) on the addition of a reducing agent. Potassium permanganate is standardized against pure oxalic acid. Oxalic acid got oxidized to the $CO_2$ by $KMnO_4$, which itself reduced to the $MnSO_4$. Oxalic acid reacts with the potassium permanganate in the following way:

Chemical reaction at room temperature is represented here as

Reduction half reaction:

$$2KMnO_4+3H2SO_4 \rightarrow K_2SO_4+2MnSO_4+3H_2O+5[O]$$

Oxidation half reaction:

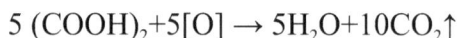

$$5(COOH)_2+5[O] \rightarrow 5H_2O+10CO_2\uparrow$$

The overall reaction involved in the process is

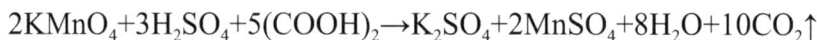

$$2KMnO_4+3H_2SO_4+5(COOH)_2 \rightarrow K_2SO_4+2MnSO_4+8H_2O+10CO_2\uparrow$$

This reaction cannot be carried on the acids like nitric acids and hydrochloric acids, as these acids chemically react with the $KMnO_4$ solution resulting in chlorine formation, which is an oxidizing agent.

Material required to carry out the experiment is as follows:

Oxalic acid, potassium permanganate, 1.0M sulfuric acid, burette, burette stand, conical flask, funnel, measuring cylinder, measuring flask, white tile, burnet, wire gauze.

**Apparatus setup:**

1. In burette- $KMnO_4$ solution

2. In conical flask- 10ml of oxalic acid + sulfuric acid

3. Indicator- self indicator ($KMnO_4$)

4. Endpoint – appearance of permanent pale pink colour.

**Procedure:** Firstly, calculate the quantity of the oxalic acid required for the 0.1N solution of potassium permanganate.

Equivalent weight of oxalic acid = molecular weight/ no of electron lost by one molecule

Equivalent weight of oxalic acid= 126/2 = 63

$$strength = normality \times equivalent\ weight$$

$$strength = \frac{1}{10} \times 63 = \frac{6.3g}{l}$$

For preparing 1L of N/10 oxalic acid solution (0.1N) amount of oxalic acid required = 6.3 g. Now, weight 6.3 g of oxalic acid in the watch glass and transfer to the measuring flask with the help of a funnel. Further, wash the funnel with the help of distilled water and make the solution upto the marked point with the help of distilled water.

**Procedure for Titration:**

1. Rinse the burette with potassium permanganate solution and fill the burette with it and fix it on burette stand and place white tile below the burette in order to find the end point correctly.

2. Pipette out 10ml of 0.1N standard oxalic acid solution in the conical flask.

3. Add a full test tube of sulfuric acid in order to prevent the oxidation of manganese.

4. Before titration with $KMnO_4$, heat the mixture up to 60°C.

5. Now note down the initial reading in the burette carefully.

6. Now titrate the hot solution against potassium permanganate solution and instantaneously swirl the solution in the flask gently.

7. Initially, the purple colour of $KMnO_4$ is cleared with oxalic acid. The appearance of permanent pink colour divulges the endpoint.

8. Repeat the titration until conformable values are obtained.

Fig. (**3**). Representation of titration of Oxalic acid with $KMnO_4$

**Applications of the Redox Titrations**

*Use of Redox Titrimetry in the Inorganic Analysis*

<u>*Chlorination of Public Water Supply*</u>

Redox titrimetry has numerous applications in the environmental, industrial and in public health analyses. One of the most imperative applications of redox titration is assessing the chlorination of water supplied to the public. Generally, chlorination of public water supplies resulted in the formation of several chlorine-containing species, collectively called total chlorine residual. Chlorine may exist in plenty of states like $Cl_2$, $OCl^-$, $HOCl$, $NH_2Cl_2$ and $NCl_3$ as free and combined chlorine residuals, respectively and the total chlorine residuals has been determined by using oxidizing power of chlorine to convert $I^-$ and $I^{3-}$. The produced amount of $I^{3-}$ has been determined by the redox titration (back titration) using $S_2O_3^-$ as titrant and starch as indicator [4].

The effectiveness of chlorination is based on chlorinating species, whether free chlorine residual or combined chlorine residuals. Later, it has been formed *via* the reaction of ammonia with free chlorine residuals. When iodide free chlorinated water sample is added to the N, N-diethyl-*p*-phenylenediamine (DPD) indicator, the free chlorine oxidizes a small part of DPD to red coloured form and then the oxidized DPD is back titrated with ferrous ammonium sulfate to its colourless form. Addition of small quantity of KI reduces the formation of monochloramine, $I_3^-$, $NH_2Cl$. The formed $I_3^-$ oxidizes DPD to the red coloured form. The above described method is also used for determining the chlorine demand for water supply. The chlorine demand is elaborated as the quantity of chlorine that must be added to a water supply in order to entirely react with any substance that can be oxidized by chlorine while also maintaining the desired chlorine residuals.

### *Determining Dissolved Oxygen in Water*

Determining the level of dissolved oxygen is very important for two reasons: firstly, for the biological oxidation of the organic and inorganic pollutants, it is a more easily available oxidant, which is vital for the care of marine life and secondly, in wastewater treatment plants. For the waste material aerobic oxidation, control of dissolved oxygen is very important. Otherwise, at a lower level of dissolved oxygen, the aerobic bacteria's are replaced by the anaerobic bacteria's that result in the formation of undesirable gases $CH_4$ and $H_2S$.

One most popular method for determining dissolved oxygen levels is Winkler method. A sample is collected in such a way that exposure to the atmosphere is prevented and then the sample is treated with $MnSO_4$ solution followed by solution of NaOH and KI. $Mn^{2+}$ is oxidized to the $MnO_2$ by the dissolved oxygen under alkaline conditions.

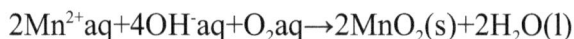

$$2Mn^{2+}aq+4OH^-aq+O_2aq \rightarrow 2MnO_2(s)+2H_2O(l)$$

After the completion of the reaction, the reaction mixture was acidified with $H_2SO_4$ and under such acidic conditions, $I^-$ got oxidized to the $I^{3-}$ by $MnO_2$.

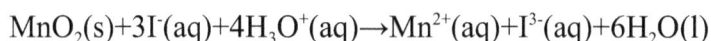

$$MnO_2(s)+3I^-(aq)+4H_3O^+(aq) \rightarrow Mn^{2+}(aq)+I^{3-}(aq)+6H_2O(l)$$

The formed $I^{3-}$ is back titrated with the $S_2O_3^{2-}$, using starch as an indicator. This method undergoes numerous interventions and some modifications have been done in order to reduce these. Some interferences like $NO^{2-}$ reduce the formation of $I^{3-}$ to $I^-$ under acidic conditions, which can be eliminated by adding sodium azide, $NaN_3$. Other interference includes the presence of reducing agents like $Fe^{2+}$, which can be eliminated by pre-treating the sample with $KMnO_4$.

## *Determination of Water in Nonaqueous Solvents*

For this titration, Karl Fischer reagent consisting of a mixture of iodine, pyridine, methanol and sulphur dioxide is used as the titrant. Pyridine is present in large concentration so that $I_2$ and $SO_2$ can be complexed with the pyridine (py) as $py.I_2$ and $py.SO_2$. When added to the water containing sample, $I_2$ is reduced to the $I^-$ and $SO_2$ is oxidized to $SO_3$.

$$py.I^2 + py.SO_2 + py + H_2O \rightarrow 2py.HI + py.SO_3$$

Methanol further prevented the reaction of $py.SO_3$ with water. The end point of the titration change from yellow colour of the products to the brown colour of Karl Fischer reagent.

### *Use of Redox Titrations in Organic Analysis*

This can be used for the determination of chemical oxygen demand (COD) in wastewater as well as in natural waters. COD measures the amount of oxygen required to completely oxidize all the organic material to the $CO_2$ and $H_2O$. This method always calculates the true oxygen demand because it is not applicable for the organic matter which cannot decompose or the matter which decompose slowly. This method is very valuable in managing industrial water-waste, as it is used to check the release of organic rich waste into the sewer system or in the environment.

The COD can be determined by refluxing the sample with excess $K_2Cr_2O_7$. Then $H_2SO_4$ is added to acidify solution, followed by the addition of $Ag_2SO_4$ as a catalyst. Mercuric sulfate ($HgSO_4$) is added to complex any chloride and to prevent precipitation of $Ag^+$ catalyst as AgCl. Refluxing was done for 2 hrs followed by cooling of solution at room temperature. The excess $Cr_2O_7^{2-}$ is back titrated with ferrous ammonium sulfate as titrant and ferroin as an indicator. Meanwhile, removing all the organic matter traces is difficult, so a blank titration is performed. The difference in the amount of ferrous ammonium sulfate is required to titrate the blank and the sample is proportional to the COD.

### *Quantification of Lithium*

Lithium serves a supreme role for high energy-density storage applications and is mandatory for the continued advancement of the world economy. However, worldwide supply of lithium is totally dependent on the lithium deposit in the selected locations, marking randomness on price and concerns for the continued production of resources. Its future demand for its use in the small electronics, renewal energy industries and automotive led to place strain on the lithium supply. Thus, recycling lithium batteries is very crucial to meet the heavy requirement for lithium-based technology. Recently, redox titration method has been developed to quantify the active lithium from the associated compounds. The method included the reduction of triiodide *via* lithium metal in a propylene carbonate or acetonitrile solvent system. The viability of the proposed redox titration method for the reduction of $I^{3-}$ by unpassivated Li(m) remaining in a used lithium-ion cell was explored. The Li(m) sample was added in excess to a 9.75 · $10^{-5}$ M $I^{3-}$ in anhydrous PC solution in an Ar purged glovebox to minimize potential contaminants to the redox reaction between $I^{3-}$ and Li(m). UV/vis spectrometry was used to track the reduction of $I^{3-}$ [5].

1. Determining reduction potential of sHdrA flavin from *Hyphomicrobium denitrificans:*

Many bacteria and archaea hired a pathway of sulphur oxidation, including an enzyme complex that is associated with the heterodisulfide reductase (Hdr or HdrABC) of methanogens. For characterization of Hdr-like proteins from sulphur oxidizers (sHdr), structural analysis of the recombinant sHdrA proteins from alpha proteobacterium *Hyphomicrobium dentrificans* at 1.4 A resolution was made. Redox titration method was developed to monitor the redox potential of sHdrA flavin by recording UV/vis spectra. As-isolated sHdrA (50 μM) was titrated stepwise with sodium dithionite, which has a redox potential below −550 mV. Stepwise addition of 2,6-dichloroindophenol (E0 = +330 mV at pH 6.0) as oxidant resulted in the complete oxidation. This experiment allowed the assignment of the redox potentials of the FAD cofactor. The ratio of oxidized/reduced FAD species was calculated from the UV-visible spectra at 456 nm and 610 nm, respectively, and plotted against the observed potential. A stable FADH• species appeared between −104 to −70 mV, recognizable by a plateau in the titration curve at 456 nm. The FAD absorption at 590–650 nm was highest at −104 mV and decreased again upon further reduction or oxidation, respectively. Titration curves can be explained by the Nernst equation [6].

2. Use of redox titration for the visual determination of alcohol content in whiskey bottles:

Recently, a foldable paper-based analytical device (PADs) has been developed to perform redox titrations in order to know the alcohol contents in the whiskey bottles. For this purpose, a classical permanganometry reaction was used based on the reaction of oxalic acid with excess permanganate in acidic media and endpoint of redox titration in different alcoholic concentrations was measured. A report on alcohol content in 44 whiskey bottles was developed by the Brazilian Federal Police and revealed that about 73% of seized samples of whiskey contained adulterated alcoholic contents. This method is very simple and of low cost, which requires short analysis time and is quite attractive for on-site forensic applications [7].

## Conclusion

This chapter basically covered the essential oxidation–reduction reactions along with the rules for assigning oxidation numbers in order to assign the movement of electrons from the species that are oxidized (reducing agents) to the species that are reduced (oxidizing agents). Added by, the sequence of steps involved in balancing half-reactions, redox titrations, and disproportionation reactions have also been covered. Various types of redox indicators like internal indicator, external indicator and self-indicator have been explained very well. Concerning, the application of redox reactions in chlorination of public water supply, quantification of lithium, analysis of alcohol contents, *etc.* has been elaborated in the current chapter. Also, oxidation–reduction reactions are often used for energy transfer in biological systems, and any deficiencies in such systems are profoundly deleterious (such as metabolic, mitochondrial, and immunologic diseases).

## CONSENT FOR PUBLICATION

Not applicable.

## CONFLICT OF INTEREST

The author declares no conflict of interest, financial or otherwise.

## ACKNOWLEDGEMENTS

Declared none.

## REFERENCES

[1]    Kolthoff, I.M. Analytical chemistry in the USA in the first quarter of this century. *Anal. Chem.,* **1977,** *49*(6), 480A-487A.
[http://dx.doi.org/10.1021/ac50014a715]

[2]    Khalid, M.A.A. *Redox: Principles and Advanced Applications*; BoD–Books on Demand, **2017.**
[http://dx.doi.org/10.5772/66005]

[3]    Stuart, S. Redox indicators. In: *Characteristics and applications*; Elsevier, **2013.**

[4]     Harvey, D. *Modern analytical chemistry. McGraw-Hill New York,* **2000**, 1.

[5]     Hebert, J.J. Quantification of Lithium *via* Redox Titration and pH Titration–A Method Comparison. **2020**.

[6]     Ernst, C.; Kayastha, K.; Koch, T.; Venceslau, S.S.; Pereira, I.A.C.; Demmer, U.; Ermler, U.; Dahl, C. Structural    and    spectroscopic    characterization    of    a    HdrA-like    subunit    from Hyphomicrobiumdenitrificans. *FEBS J.,* **2021**, *288*(5), 1664-1678.
        [http://dx.doi.org/10.1111/febs.15505] [PMID: 32750208]

[7]     Nogueira, S.A.; Lemes, A.D.; Chagas, A.C.; Vieira, M.L.; Talhavini, M.; Morais, P.A.O.; Coltro, W.K.T. Redox titration on foldable paper-based analytical devices for the visual determination of alcohol content in whiskey samples. *Talanta,* **2019**, *194*, 363-369.
        [http://dx.doi.org/10.1016/j.talanta.2018.10.036] [PMID: 30609544]

# CHAPTER 6

# Complexometric Titrations

**Anju Goyal**[1], **Ramninder Kaur**[1,*], **Sandeep Arora**[1] and **Komalpreet Kaur**[2]

[1] *Chitkara College of Pharmacy, Chitkara University, Rajpura*

[2] *G.H.G Khalsa College of Pharmacy, Gurusar Sadhar, Ludhiana*

**Abstract:** Complexometric titrations (also known as chelatometry) are used mainly to determine metal ions by use of complex-forming reactions. It is a form of volumetric analysis in which the formation of a colored complex is used to indicate the endpoint of a titration. In this method, a simple ion is transformed into a complex ion and the equivalence point is determined using metal indicators or electrometrically. An indicator capable of producing an unambiguous color change is usually used to detect the endpoint of the titration. The versatility, sensitivity, and general convenience of complexometric titrations are dependent on the correct choice of indicators endpoint detection.

**Keywords:** Complexation, Metal ions, Metal ion indicators, Titrations.

## INTRODUCTION

Complexometric titration is one in which a soluble, undissociated and stoichiometric complex is formed during the addition of titrant to the sample solution (usually solution of a metal ion). It is the method of volumetric analysis developed after the introduction of the analytical reagent commonly known as an ethylene diamine tetraacetic acid with di-sodium salt [EDTA] [1].

Complexometric titrations are those reactions in which simple metal ions are transformed into of complex by the addition of a reagent which is known as ligand and complexing agent. The complex formed is stable and water soluble complexes are called sequestering agents [2].

## COMPLEXATION

The process of complex ion formation is known as complexation.

* **Corresponding Author Ramninder Kaur**: Chitkara College of Pharmacy, Chitkara University, Rajpura; Email: ramninder@gmail.com

**Anju Goyal & Harish Kumar (Eds.)**
**All rights reserved-© 2022 Bentham Science Publishers**

A Complexometric reaction with metal ion involves the replacement of one or more co-ordinate solvent molecules by other nucleophilic groups. These groups bind to the central metal ion known as ligands and in aqueous solution, the reaction can be represented as:-

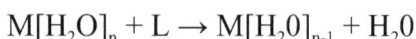

$$M[H_2O]_n + L \rightarrow M[H_2O]_{n-1} + H_2O$$

n = coordination number

M = metal ion

In this reaction there is successive replacement of water molecule by ligands groups until the complex is formed. The vast majority of Complexometric titrations are carried out using multidentate ligands such as EDTA [1].

## CHELATE

It is the complex that is formed by the combination of polyvalent metal ions with a molecule which essentially contains two or more groups that can donate electrons. These rings are generally more water soluble chelates. They are the sequestering agents [1].

Examples of chelating agents:

- **8-hydroxy quinoline**
- **Dimethyleneglyonime**
- **Salicyladoxine**
- **EDTA**

## LIGANDS

Complexating agents in any electron donation ion or molecule are called as ligands that have ability to form one or more covalent bond with the metal ion. They can be any electron donating group that has the ability to bind with the metal ion producing a complexion ion [2].

### Classification of Ligands

### *Monodentate Ligand*

It is the ligand that is bound to the metal ion at only one point by the donation of a lone pair of electrons to the metal. *i.e.* which possess only one electronegative donor atom (O, N, C or Cl$^-$) containing only one unshared pair of electrons. Example: - ammonia which forms complex with cupric ion.

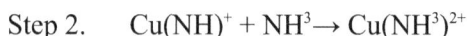

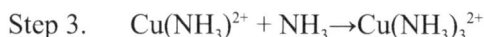

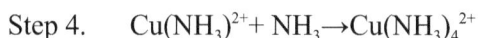

$$\text{Step 1.} \quad Cu^{2+} + NH_3 \rightarrow Cu(NH)^{2+}$$

$$\text{Step 2.} \quad Cu(NH)^+ + NH^3 \rightarrow Cu(NH^3)^{2+}$$

$$\text{Step 3.} \quad Cu(NH_3)^{2+} + NH_3 \rightarrow Cu(NH_3)_3^{2+}$$

$$\text{Step 4.} \quad Cu(NH_3)^{2+} + NH_3 \rightarrow Cu(NH_3)_4^{2+}$$

Simple ligands, such as halide ions or the molecules such as $H_2O$ or $NH_3$, $CN^-$ $NO^{2-}$ and Cl$^-$ are Monodentate.

### Bidentate and Multidentate Ligands

These ligands are known to contain more than one group and are capable of binding with metal ion. They include bidentate ligands (2 donor atoms), tridentate ligand (3 donor), and quadridentate ligands *e.g.*, bis (ethylenediamine) cobalt III. In other words, when the ligand molecule or ion has two atoms, each of which has a lone pair of electrons, then the molecule has two donor atoms and it may be possible to for two-coordinate bonds with the same metal ion, which is called a bidentate ligand.

Example: - tris (ethylenediamine) Cobalt (III) complex. [Co (en) 3]$^{3+}$

In this six co-ordinate octahedral complex of cobalt (III), each Bidentate ethylenediamine (1, 2-diainoethane) molecule is bound to the metal ion through the lone pair electrons of the two nitrogen atoms. These result in the formation of the 3 five-member rings, each including the metal ion; the process is called chelation. It can be hexadentate.

EDTA or 1, 2-diaminoethanetetraacetic acid or 1, 2- [Bis (carboxymethyl) amino] ethane. *i.e.* containing 2, 3,4,5,6, *etc.* donor atom within the same molecule. All the polydentate ligands are called chelating ligands [1].

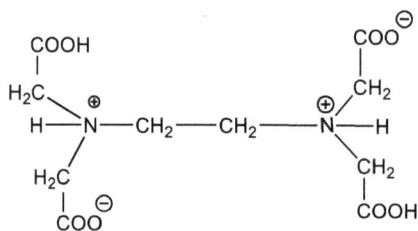

EDTA

Hexadentate ligand which may coordinate with a metal ion through its two N-atoms and 4-Carbonyl groups.

## CO-ORDINATE COVALENT BONDS OR DATIVE BONDS

In this bond formation, electron pair is required and it is contributed by the atom only. Atom contributing to the electron pair is called as donor, and another atom that shares the electron pair is called acceptor [3].

$$H_3N + BF_3 \rightarrow [H_3 BF_3]$$

## COVALENT BOND

The bond is formed by the mutual sharing of electron pairs called covalent bond and this covalent bond is formed between atoms of the same element or different elements. Each shared electron pair is called bond pair and is represented by a line between the atoms [1].

## WERNER'S COORDINATE NUMBER (WCN)

It is the number of small groups which are attached to the central atom in a complex. It is not related to the valency of the complex having small groups, but it is related to the space available for the attachment. Elements can be attached and accommodated in a particular way. *Elements of the second period can accommodate 4 groups in such a way [4].

- Third period elements can accommodate 6 groups.
- Fourth period can accommodate 8 groups at a time.

And this WCN is required to know the maximum no. of small groups which can be accommodated around a central ion.

## STABILITY CONSTANT

The stability of complex ion varies within very wide limits. It is quantitively expressed by means of the stability constant. The more stable complex, greater the stability constant *i.e.*, the smaller the tendency of the complex ion to dissociate into its constituent ions [2].

## STABILITY

The thermodynamic stability of a species is a measure of the extent to which this species will be formed from other species under certain conditions.

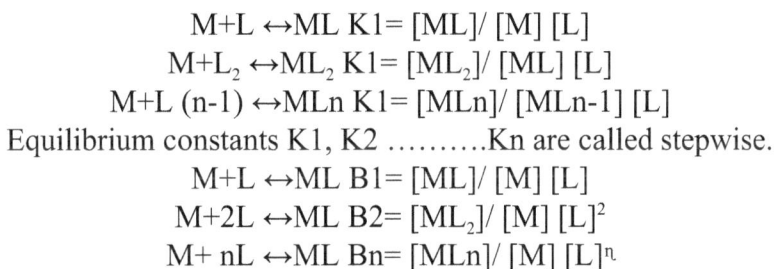

$$M+L \leftrightarrow ML \quad K1= [ML]/ [M] [L]$$
$$M+L_2 \leftrightarrow ML_2 \quad K1= [ML_2]/ [ML] [L]$$
$$M+L (n-1) \leftrightarrow MLn \quad K1= [MLn]/ [MLn-1] [L]$$

Equilibrium constants K1, K2 ..........Kn are called stepwise.

$$M+L \leftrightarrow ML \quad B1= [ML]/ [M] [L]$$
$$M+2L \leftrightarrow ML \quad B2= [ML_2]/ [M] [L]^2$$
$$M+ nL \leftrightarrow ML \quad Bn= [MLn]/ [M] [L]^n$$

The equilibrium constants B1, B2 ............... Bn are called the overall stability constants.

The stability of the complex can be altered *i.e.* it can be decreased or increased.

- Stability of the metal EDTA complexes may be altered by reaction.
- By the variation in pH.
- By the presence of other complexing agents.
- Stability can be decreased by increasing temperature and acidic pH.
- Stability can be increased by the addition of Ethanol.

- If the stability of metal complexes alters, then the stability is called Apparent or Conditional stability constant.
- The buffers are used to form the most stable complexes.

*E.g.,* Ammonium chloride solution

## TYPES OF COMPLEXOMETRIC TITRATIONS

In the EDTA titration, if pM (negative logarithm of the free metal ion conc., pM= -Log [$M^{n+}$]) is plotted against the volume of EDTA solution added, a plotted against the volume of EDTA solution added, a point of inflexion occurs at the equivalence point in some instances this sudden increased may exceed 10 pm units [1].

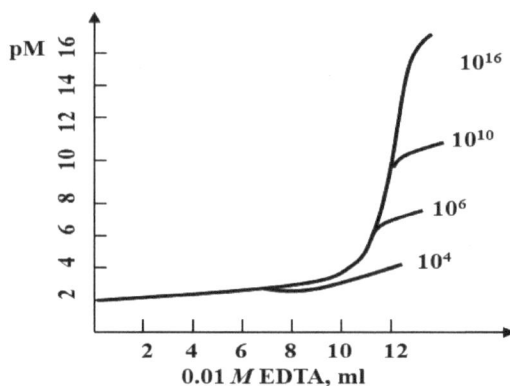

**Fig. (1).** This figure shows the general shape of titration curves obtained by titrating 10.0ml of a 0.01M solution of a metal ion M with a 0.01M EDTA solution.

The apparent stability constant of various metal EDTA complexes are indicated at the extreme light of the curves.

In EDTA titration, a metal ion sensitive indicator, metal indicator or metal ion indicator is often employed to detect the changes of pH. Such indicators for complexes with specific metal ion which change at the equivalent points.

## TYPES OF TITRATIONS:

1. Direct titration

2. Indirect or back titration

3. Replacement and substitution titration

4. Alkalimetric titration

### Direct Titration

It is a simple and most convenient method in EDTA titrations. The solution containing the metallic salts is buffered to the desired pH and directly titrated with EDTA standard solution by using suitable pH indicators until the color change is absorbed. It may be necessary to prevent precipitation of the hydroxide of the metal by addition of sum auxiliary Complexometric agents. For example: tartrate, citrate, triethanolamine. A blank titration may be performed by omitting the sample (metallic salts) as the check on the metallic impurities present on the reagent. Metal ions such as calcium, magnesium, zinc directly estimated by direct titration method. For example: magnesium sulphate, magnesium trisilicate, zinc [2].

### Indirect or Back Titration

Many metals like aluminium form hydroxides for various reasons and they cannot be titrated by directed method. They may precipitate from the solution or they may inert complexes or suitable indicator is not available.

In such cases, excess of standard EDTA solution is used to the resulting and it is back titrated with a standard metal ion solution such as $ZnCl_2$, $MgCl_2$, or help of pH indicator [2].

Example: Aluminium sulphate, Aluminium glycinate, Bismuth sub carbonate and dried aluminium hydroxide.

The indirect titration is necessary for following cases: -

1. For insoluble substances like lead sulfate, calcium oxalate.

2. Reactions occurring slowly with EDTA.

3. Those metabolisms which form a more stable complex with EDTA than with the desired indicator.

4. For those metals which precipitate as hydroxides for solution at the pH are required for titrations.

In this titration, excess of standard EDTA solutions and suitable buffer solution is added to the metal solution or suspension. The solution is heated to effect complex formation, cooled and the EDTA not consumed by the sample, Back titration with Mg and Zn chloride using a suitable indicator [1].

## Replacement or Substitution Titration

These titrations may be used for metal ions that do not react with an indicator or form EDTA complex that is more stable than other metals such as Mg and Ca. In these of titration, the metal ion may be determined by the displacement of an equivalent amount of Mg or Zn from a less stable edetate complex [1].

## Alkalimetric Titration

When a solution of disodium edetate is added to a solution containing metallic and the complexes are formed with the liberation of two equivalents of hydrogen ions then these liberated hydrogen ions are displaced by heavy metal and can be titrated with a standard solution hydroxide using an acid-base indicator [1].

Alternatively, an iodate-iodide mixture is added to the EDTA solution and liberated iodide is titrated with standard thiosulphate solution. These types of titrations are carried out in unbuffered solution.

## MASKING AND DEMASKING AGENTS

### Masking Agents

Masking may be defined as the process in which a substance without physical separation of it or its reaction products is so transformed that it does not enter into a particular reaction [3].

By using masking reagent, some of the cations in a mixture can often be masked so they can no longer react with EDTA or with the indicator.

Example: ammonium fluoride, Ascorbic acid, Dimercaprol, Potassium iodide.

### Need

When it is required to estimate selectively one or more ions in a mixture of cations and to eliminate the effects of possible impurities which would add to the titrate; "Masking agents" are used.

## Examples of Masking Agents

Cyanide ion effective masking agent. It forms stable cyanide complexes with various cations. For example: Cd, Zn, Hg, Co, Ag, Ni metals so, it is, therefore, possible to determine cation such as Ca, Mg, Pb and manganese. In the presence of these metals by masking with a potassium or sodium cyanide.

## Demasking Agents

It is the process in which the mask substance regains its ability to enter into a particular reaction [3].

Example: formaldehyde or Chlorhydrate can be used to release the mask zinc ions by potassium cyanide.

Masking and demasking can be done by one of the three methods: -

1.Addition of precipitants.

2.Addition of complexing agents.

3. pH control

## METAL ION INDICATORS USED IN COMPLEXOMETRIC TITRATIONS

These are also known as metal ion indicator or pM indicator. Metal ion indicator is a dye that forms a complex with metal that have one color when it binds with metal ions, but it may have different color when it is not binds to the metal ions.

### The metal ion indicators should comply with the following requirements:

- The compound should be chemically stable throughout the titration.
- It should form 1:1 complex weaker than the metal Chelate complex.
- Color of the indicator and the metal complexed indicator should be different.
- Color reaction should be selective for the metal being titrated.
- The indicator should not compete with the EDTA.

## COMMON EXAMPLES

## Mordant Black-II

Titratable metal ions: - Mg, Zn, Cd, Hg, Pb, Ca, Ba

Color change: - red to blue

pH range:- 6-7

It is blue at about pH-10 and most of the complexes are reddish below pH-6.3 and about pH-11.5, it gives reddish color [4].

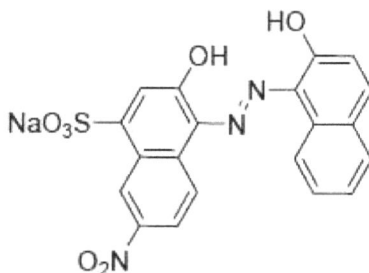

## MUREXIDE (AMMONIUM PURPURATE)

It was the first metal ion indicator to be implied in the EDTA titration.

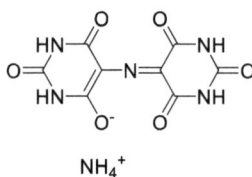

Description: - up to pH-9 it gives reddish violet color between pH-9-11 gives violet color

Above pH-11 gives blue violet color.

Titratable metal ions: - Cu, Co, Ni, and lanthanoids.

pH range:- 12

Color change: - violet to blue.

## XYLENOL ORANGE (LEMON TO YELLOW)

It is an acid base indicator. Yellow color change in acid solution and of red color in alkaline solution.

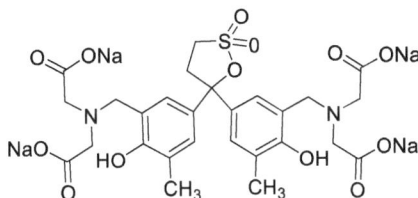

Titratable metal ions: - Bi, thorium, Pb, Cd, Hg

Description: - Its pH range varies from 4-7. It gives violet color with Hg, Pb and Zn in alkaline solution and gives yellow color in EDTA solution.

## SOLOCHROME DARK BLUE OR MORDANT BLACK-17 OR CALCON

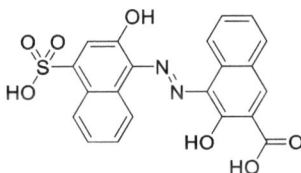

Titratable metal ions: - mainly used for assay.

For example: CaCl, calcium gluconate and calcium and sodium lactate.

Description: - gives reddish purple color with Ca in alkaline solution and blue color in the absence of free ca+ ions.

## ALIZARIN FLUORINE COMPLEX OR ALIZARINE FLUORINE BLUE OR ALIZARINE COMPLEX ONE

Color change: red to yellow

pH range:- 4-3

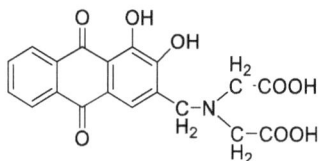

Titrable ions; Pb, Zn, Co, Hg,

Description: it may be used in acid solution at pH 4-3 and color change from red to yellow.

## Other examples:

1. Catechol violet: (pH range: 8-10)

Metals detected:Mn, Mg, Fe, Co,Pb

2. Calmagite: (pH range: 0-12)

Metals detected: Mg,Ca,Co,Zn,Cd,Mn

3. Methyl thymol blue: (rare earth metals).

## ASSAY OF MAGNESIUM SULPHATE

**Description:** It is colorless, crystalline powder, and has colorless crystals or white powder [5]. It contains not less than 99% or not more than 100.5% of $MgSO_4$.

Molecular formula: $MgSO_4.7H_2O$

Molecular weight: 246.5 gm

## PREPARATION AND STANDARDIZATION OF 0.05M EDTA SOLUTION

18.6g of disodium EDTA and make up the volume 1000ml with distilled water.

## STANDARDISATION

0.8g granulated zinc +warm 12ml dilute HCl +0.1m bromine water. Now boil the solution.

↓

Cool the solution and make up the volume 250ml.

↓

Then take 25ml of this solution into a conical flask and neutralize it with 2M NaOH solution.

↓

Dilute it to about 150ml with water and add water and add sufficient $NH_3$ buffer having pH 10 to dissolve the precipitate.

↓

Add 5ml in excess and add pM indicator mordant black-II (50mg)

↓

Titrate with disodium edetate solution until the solution turns green.

**ASSAY:** To 0.3g $MgSO_4$+50ml distilled water add 10ml strong ammonia ammonium chloride. Start the titration with EDTA, then add 0.1g Mordant black-II indicator.

Endpoint is blue colour.

## ASSAY OF ALUMINIUM HYDROXIDE GEL

**Description:** A white viscous suspension, translucent in thin layer. It contains not less than 3.5% and not more than 4.4% $^w/_w$ of aluminium hydroxide.

Molecular formula: $Al_2 (OH)_3$

**Procedure for Assay:**

Weigh 5g sample +3ml HCl warming on a water bath.

↓

Cool to below 20°C and dilute to 100ml with distilled water

↓

Then take 20ml from this solution+ add 40ml 0.05M EDTA+ 80ml distilled water + 0.15ml methyl red solution

↓

Neutralized it with 1M NaOH then warm on a water bath for 30 minutes, add 3ml hexamine Titrate

↓

With 0.05M lead nitrate using 0.5ml xylenol orange as indicator.

↓

Endpoint is blue-green colour.

## ASSAY OF CALCIUM GLUCONATE

Molecular formula: $C_7H_{22}CaO_{14}.H_2O$

Molecular weight: 448g

### Description:

1. It is white crystalline powder or granules.

2. It contains not less 98.5% or not more than 102% of calcium gluconate.

### Procedure:

0.8g of sample dissolved in 150ml distilled water then add 5ml 0.05M $MgSO_4$+10ml strong $NH_3$ solution.

↓

Titrate with .05M disodium edentate using mordant black –II as indicator.

↓

End point-deep blue colour and at the end point from the volume of 0.05M disodium edentate required the subtraction from the volume of the $MgSO_4$ solution added.

## CONCLUSION

Complexometric titrations (also known as chelatometry) are used mainly to determine metal ions by the use of complex-forming reactions, resulting in the formation of colored complex indicating end point of the reaction. In these types of reactions, simple metal ions are transformed into of complex by the addition of a reagent which is known as ligand and complexing agent. Various types of ligands like monodentate ligands, bidentate and multidentate ligands have been elaborated with the help of examples. Wide varieties of indicators used for these types of titrations have been listed along with their chemical structure, indicating endpoint of the reactions. Different types of titrations like direct titrations, indirect titrations, substitutions and replacement titrations and alkalimetric titrations have been explained very well, using examples and procedures.

## CONSENT FOR PUBLICATION

Not applicable.

## CONFLICT OF INTEREST

The author declares no conflict of interest, financial or otherwise.

## ACKNOWLEDGEMENTS

Declared none.

## REFERENCES

[1]     Jeffery, G.H.; Bassett, J.; Mendham, J.; Denny, R.C. *Vogel's Text Book of Quantitative Chemical Analysis: Complexation Titrations,* 5th ed; Longman Scientific & Technical, **1989**.

[2]     Svehla, G. *Vogel's Text Book of Inorganic Macro and Semimicro Qualitative Inorganic Analysis: Complexation Reactions,* 5th ed; Longman Group Limited, **1979**.

[3]     Bentley, A.O.; Driver, J.E. *Textbook of Pharmaceutical Chemistry: Organometallic Complexes in Analysis,* 7th ed; Oxford University Press, **1977**.

[4]     Kennedy, J.H. *Analytical Chemistry: Principles: Complexation Titrations,* 2nd ed; Saunders College Publishing: New York, **1990**.

[5]     Beckett, A.H.; Stenlake, J.B. *Practical Pharmaceutical Chemistry.,* **1975**, *Vol. I and II.*

# CHAPTER 7

# Diazotization Method

**Kamya Goyal[1,2,*], Navdeep Singh[1], Shammy Jindal[1], Rajwinder Kaur[2], Anju Goyal[2] and Rajendra Awasthi[3]**

[1] *Laureate Institute of Pharmacy, Jawalamukhi, Himachal Pradesh, India*

[2] *Chitkara College of Pharmacy, Chitkara University, Rajpura, Punjab, India*

[3] *Amity Institute of Pharmacy, Amity University Uttar Pradesh, Noida, Uttar Pradesh, India*

**Abstract:** This method was first invented in the year 1858 and after the proper research, this method was easy to be applied in synthetic dye industry. This method was first used in the determination of dyes. The progression of reaction and the mechanism concerned in this method was first predictable by scientist Peter Griess. The major mechanism of this method is when the aromatic amine which is primary in nature essentially starts a reaction with acidic environment consisting of sodium nitrite, which is further changed into diazonium salt. A diazonium salt is fashioned which is close with one other amine (phenol), and distorted into intense colored azo dye. The basic principle concerned in this method is based on the amines which are primary aromatic in nature, after that the sample reacts with sodium nitrite compound in the occurrence of HCl hydrochloric acid to form a diazonium salt. The end point is detected by the formation of the blue colour with starch iodide paper. Another method for the detection of end point is by immersing the platinum electrodes in the resulting solution and it is also detected by the dead-stop end point method. The different titration method includes direct method, indirect method and other method used in Diazotization technique. In the procedure, a potentiometer and electrode system were used, and the deflection is observed. The temperature of the reaction was maintained at 0-5 °C. So, in this method we studied the basic principle involved, procedure, methods of titration, and the appropriate applications of this technique.

**Keywords:** Absorbance, Applications, Azo dye, Determination of dyes, Diazonium ion, Diazotization method, Direct method, End point, External indicator, Indirect method, Nano technology, Nitrite, Peter Griess, Potentiometer, Principle, Procedure, Standard flask, Titration.

## INTRODUCTION

The diazotization method is an excellent method which is used to transform the primary aromatic amines in the form of diazonium compound. This method was

* **Corresponding Author Kamya Goyal:** Laureate Institute of Pharmacy, Jawalamukhi, Himachal Pradesh, India and Chitkara College of Pharmacy, Chitkara University, Rajpura, Punjab, India; Email: kamya.goyal7@gmail.com

**Anju Goyal & Harish Kumar (Eds.)**
**All rights reserved-© 2022 Bentham Science Publishers**

first invented in the year 1853, and after the proper invention this method was accessible to be applied in synthetic dye industry. The process of reaction and, the mechanism involved in this method was first anticipated by scientist Peter Griessin. The main mechanism of this method started when the aromatic amines which are primary in nature basically start a reaction with acidic environment consisting of sodium nitrite, which is further transformed into diazonium salt. As we discuss, this method is firstly used to determine the dyes. In this method, the nitrite compound may accurately estimate, and it was used for diazotization reaction at a very low concentration with certain aromatic compounds. A diazonium salt is formed which is attached with another amine (phenol), and changed into intense colored azo dye [1, 2].

## Principle

The conversion of nitrate to nitrite is reduced by hydrazine in an alkaline solution, since hydrazine reduces nitrate to nitrite. The nitrite compound was further diazotized with sulfanilamide and the vacant compound of diazonium was consequently joined with I-naphthylethylenediamine which is later transformed to an intense red colored dye. The absorbance was measured at 535 m« and was directly comparative to the attentiveness of nitrate compound which occurred as unique sample. The alone nitrite compound was determined in the occurrence of nitrate in a parallel procedure, apart from that the reduction step was absent [3].

The basic principle implicated in this method is based on the amines which are primary aromatic in nature, after that the sample reacts with sodium nitrite compound in the occurrence HCL hydrochloric acid to form a diazonium salt [4].

$$R-NH_2 + NaNO_2 + HCl \longrightarrow R-N^+\equiv N-Cl^- + NaCl + H_2O$$

In the solution of amine the sodium nitrites were added with the acid at 0-5 °C. After that, the amine would react with nitrous acid changed into nitrosamine followed by the tautomerization reaction [5]. The molecule of water was missing to form a diazonium compound. The diazonium ion was stabilized to form a displacement reaction with positive charge at two positions of ortho and para of the ring.

$$C_6H_5-NH_2 + NaNO_2 + HCl \longrightarrow C_6H5-N^+\equiv N-Cl^- + NaCl + H_2O$$

**Battery**

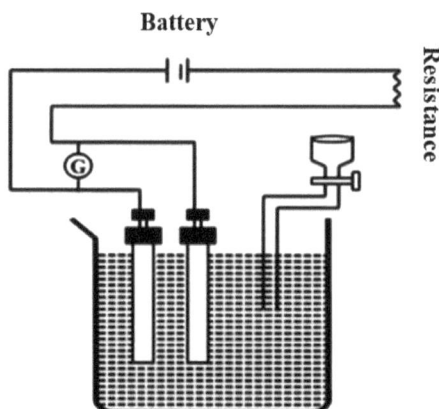

**Fig. (1).** Potentiometric technique for the determination of end point.

## Theory

When sodium nitrite is reacted with the hydrochloric acid, sodium chloride and nitrous acid are formed [6].

$$NaNO_2 + HCl \longrightarrow NaCl + HNO_2$$

The obtained nitrous acid is reacted with the primary aromatic amine in the form of diazonium salt. The glut of nitrous acid is removed by the addition of ammonium sulphate solution.

$$R-NH_2 + HNO_2 \longrightarrow R-N=NH + H_2O$$

The endpoint is detected by the formation of the blue colour with starch iodide paper. This is organized by immersing the filter paper in the starch mucilage and potassium iodide solution.

$$KI + HCl \longrightarrow KCl + HI$$

$$2HI + 2HNO_2 \longrightarrow I_2 + 2NO + 2H_2O$$

$$I_2 + starch\ mucilage \longrightarrow blue\ colour\ end\ point$$

## End Point Detection

To determine the endpoint in a diazotization method (titration process), various steps should followed:

- To determine the surplus of nitrous acid in experiment starch iodide were used as external indicator. After the complete process of diazotization, a single drop of vacant solution was added into starch iodide paper, which converts into dark colour. Another method for the detection of endpoint is by immersing the platinum electrodes in the resulting solution and it is also detected by the dead-stop endpoint method.
- Another technique used to determine the end point is diazotization titration. In this technique vigorously adding the potassium iodide with surplus of acid leads to liberates the iodine [7]. The active form of iodine is titrated again with sodium thio-sulphate. In this procedure starch was used as an external indicator. At the end, the endpoint of the reaction was detected in the form of blue colour.

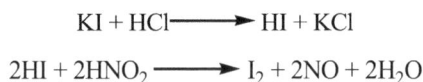

$$KI + HCl \longrightarrow HI + KCl$$
$$2HI + 2HNO_2 \longrightarrow I_2 + 2NO + 2H_2O$$

## Preparation and Standardization of the Sodium Nitrite Solution

Correctly weigh the sodium nitrite then dissolve it with water and the desired volume would make upto the mark. To standardize the sodium nitrite a sulphanilamide was dissolved with water, and conc. HCl solution. Later, the solution was cooled at 15° C with sodium nitrite which is a standard solution.

## Factors Affecting the Diazotization

1. Concentration of acid.

2. $NaNO_2$ pH.

3. Reaction temperature would maintain in the range of 0-5 °C because the diazonium compounds were decomposed when the temperature rose from this range.

4. The time of reaction will maintained in between 10–15 min due to the reaction of compound with nitrous acid at multiple rates, which is totally depend on the compound nature.

5. Slow reaction of diazotizable groups includes sulpha, carboxylic, and nitrogen oxide group.

6. Fast reaction of diazotizing groups includes anilide, toluidine and aminophenol [8].

## DIAZOTIZATION TITRATIONS METHODS

The diazotization titration experiment was mainly consisting of three methods which are explained below:

### 1. Direct Method

Direct method of diazotization was mainly depending on the principle of treating amino groups containing drugs with acid solution. The vacant solution was kept inside the cold water, and water bath with ice cubes to maintain the temperature in the range of 0–5 °C. After this process, the solution was titrated along with solution of sodium nitrite. The end point of reaction was recognized by the methods which are mentioned below [9].

### 2. Indirect Method

Indirect method was works on the principle when the surplus of nitrous acid was filled in the sample solution of titration, and with that the solution was titrated reversibly with another suitable titrant [18]. This indirect method of titration was mostly used to determine the diazonium salts which are insoluble in nature [10].

### 3. Other Method

The principle function implicated inside this method is to make diazo oxide, as this diazo component is highly stable in nature. The example includes, when the aminophenols were willingly oxidized because of nitrous acid then it will change into quinones with the occurrence of copper sulphate solution. This system of reaction produces an oxidized diazonium compound. This willingly undergoes the coupling reaction with the nitrous acid [11].

### Procedure

- Firstly the sample was weighed in required quantity, and then transferred into the standard flask.
- Concentrated forms of hydrochloric acid (HCl) with potassium bromide (KBr) were added, and the vacant volume was made up with distilled water. The consequential (resultant) solution was concluded as a standard solution.
- The suitable volume from the standard solution (resultant) was pipetted out, and

the temperature of the reaction was maintained in the range of 0-5 °C.

• After that the process of titration started along with sodium nitrite solution, the process would continuously go till the blue colour was obtained on starch iodide paper [12, 13].

Another procedure is after maintaining the conical flask temperature, a pair of platinum electrodes is immersed. Then the electrodes are connected to the potentiometer and slowly titrated with sodium nitrite solution until a permanent deflection is observed at the endpoint.

### Applications

• This method is applicable in nano technology.
• It is also used in the preparation of hydrocarbons, aryl halide, aryl cyanide, and aryl hydrazine.
• It is used in the assay of sulpha drugs including dapsone, sulphonamides, sulphacetamide sodium, sulphadiazine, sulphamethazole, sulphadoxine, sulphamethoxazole, sulphaphenazone, *etc.*
• It is used in the assay of some drugs including benzocaine, procainamide, procaine, suramin, sodium amino salicylate, primaquine sulphate, *etc.*
• This method is used in the determination of chloropheneramine, dopamine, procaine, amphetamine, procaine, ephedrine, and P-amino benzoic acid (vitamin $B_4$) [14].

### CONCLUSION

This method was easy for application in synthetic dye industry. When naturally occurring aromatic amines combine with an acidic environment to produce sodium nitrite followed by formation of diazonium salt, this approach is used. The diazotization technique is a good method for transforming primary aromatic amines into the diazonium compound. In the analytical procedure, three methodologies were used *viz.* direct, indirect, and other. To justify the reaction, the endpoint should also be established. This technology is applicable in a variety of sectors.

### CONSENT FOR PUBLICATION

Not applicable.

### CONFLICT OF INTEREST

The authors declare no conflict of interest, financial or otherwise.

# ACKNOWLEDGEMENTS

Declared none.

# REFERENCES

[1]     Mullin, J.B.; Riley, J.P. The spectrophotometric determination of nitrate in natural waters, with particular reference to sea-water. *Anal. Chim. Acta,* **1955,** *12,* 464-480.
[http://dx.doi.org/10.1016/S0003-2670(00)87865-4]

[2]     Ahmed, M.J.; Stalikas, C.D.; Tzouwara-Karayanni, S.M.; Karayannis, M.I. Simultaneous spectrophotometric determination of nitrite and nitrate by flow-injection analysis. *Talanta,* **1996,** *43*(7), 1009-1018.
[http://dx.doi.org/10.1016/0039-9140(95)01824-7] [PMID: 18966574]

[3]     Kumar, K.; Bilwa, M. A facile spectrophotometric determination of nitrite using diazotization with p-nitroaniline and coupling with acetyl acetone. *Mikrochim. Acta,* **2001,** *137*(3-4), 249-253.
[http://dx.doi.org/10.1007/s006040170018]

[4]     Cherian, T.; Narayana, B. A new system for the spectrophotometric determination of trace amounts of nitrite in environmental samples. *J. Braz. Chem. Soc.,* **2006,** *17*(3), 577-581.
[http://dx.doi.org/10.1590/S0103-50532006000300022]

[5]     Sreekumar, N.V.; Narayana, B.; Hegde, P.; Manjunatha, B.R.; Sarojini, B.K. Determination of nitrite by simple diazotization method. *Microchem. J.,* **2003,** *74*(1), 27-32.
[http://dx.doi.org/10.1016/S0026-265X(02)00093-0]

[6]     Metwally, M.E. Primaquine phosphate as a promising substitute for N-(1-naphthyl) ethylenediamine; II. Analysis of sulfa drugs in pharmaceutical dosage forms and biological samples. *Anal. Sci.,* **1999,** *15*(10), 979-984.
[http://dx.doi.org/10.2116/analsci.15.979]

[7]     Tan, F.; Cong, L.; Li, X.; Zhao, Q.; Zhao, H.; Quan, X.; Chen, J. An electrochemical sensor based on molecularly imprinted polypyrrole/graphene quantum dots composite for detection of bisphenol A in water samples. *Sens. Actuators B Chem.,* **2016,** *233,* 599-606.
[http://dx.doi.org/10.1016/j.snb.2016.04.146]

[8]     Wan, J.; Si, Y.; Li, C.; Zhang, K. Bisphenol a electrochemical sensor based on multi-walled carbon nanotubes/polythiophene/Pt nanocomposites modified electrode. *Anal. Methods,* **2016,** *8*(16), 3333-3338.
[http://dx.doi.org/10.1039/C6AY00850J]

[9]     Karimi Zarchi, M.A.; Ebrahimi, N. Diazotization□iodination of aromatic amines in water mediated by crosslinked poly (4□vinylpyridine) supported sodium nitrite. *J. Appl. Polym. Sci.,* **2011,** *121*(5), 2621-2625.
[http://dx.doi.org/10.1002/app.34004]

[10]    Joy, A.; Anim-Danso, E.; Kohn, J. Simple, rapid, and highly sensitive detection of diphosgene and triphosgene by spectrophotometric methods. *Talanta,* **2009,** *80*(1), 231-235.
[http://dx.doi.org/10.1016/j.talanta.2009.06.059] [PMID: 19782219]

[11]    Ensafi, A.A.; Amini, M.; Rezaei, B. Molecularly imprinted electrochemical aptasensor for the attomolar detection of bisphenol A. *Mikrochim. Acta,* **2018,** *185*(5), 265.
[http://dx.doi.org/10.1007/s00604-018-2810-x] [PMID: 29691660]

[12]    Ragavan, K.V.; Rastogi, N.K.; Thakur, M.S. Sensors and biosensors for analysis of bisphenol-A. *Trends Analyt. Chem.,* **2013,** *52,* 248-260.
[http://dx.doi.org/10.1016/j.trac.2013.09.006]

[13]    Butler, R.N. Diazotization of heterocyclic primary amines. *Chem. Rev.,* **1975,** *75*(2), 241-257.
[http://dx.doi.org/10.1021/cr60294a004]

[14] Narayana, B.; Sunil, K. A spectrophotometric method for the determination of nitrite and nitrate. *Eurasian J. Anal. Chem.,* **2009**, *4*(2), 204-214.

## CHAPTER 8

# Kjeldahl Method

**Kamya Goyal[1,2,\*], Navdeep Singh[1], Shammy Jindal[1], Rajwinder Kaur[2], Anju Goyal[2] and Rajendra Awasthi[3]**

[1] *Laureate Institute of Pharmacy, Jawalamukhi, Himachal Pradesh, India*

[2] *Chitkara College of Pharmacy, Chitkara University, Rajpura, Punjab, India*

[3] *Amity Institute of Pharmacy, Amity University Uttar Pradesh, Noida, Uttar Pradesh, India*

**Abstract:** This method was introduced by Johan Kjeldahl in 1883 for the quantitative estimation of nitrogen in a compound which becomes a classical and widely employed method in analytical chemistry and has been extensively utilized from over more than 130 years. The presence of nitrogen in organic compounds and in other protein materials was identified by this technique. The amount of protein was calculated from the different varieties of material including food for human beings, some fertilizers, fossil fuels and other water waste. The process in this method to oxidize the compounds containing carbon dioxide or hydrogen atoms are changed in water. The ammonium ions are further transformed into ammonia gas when they dissolve in oxidized solution. The three step principle was describing the process of nitrogen estimation, in which the first step is digestion, which deals with the conversion of nitrogen in the food material into ammonia. In second step, the process of neutralization started, in which ammonium sulphate changed into ammonia gas. The third last step of titration was started and a suitable indicator was used to detect the end point of reaction. The first step in the procedure of this method is digestion and this process was done for at least 60-90 minutes. The second step deals with the distillation process in which the nitrogen is separated and the third step deals with the titration in the presence of acid and this will give us the determination of ammonia compound in the sample. In this chapter, we highlight the basic fundamentals, principle, procedure, applications and also recent advancements were covered.

**Keywords:** Ammonia, Applications, Catalyst, Digestion Flask, Distillation, Fertilizers, Fossil fuels, Human food, Hydrogen ion concentration, Indicator, Johan Kjeldahl, Kjeldahl method, Neutralization, Nitrogen Estimation, Principle, Procedure, Protein, Recent advancements, Titration, Waste water.

---

**\* Corresponding Author Kamya Goyal:** Laureate Institute of Pharmacy, Jawalamukhi, Himachal Pradesh, India and Chitkara College of Pharmacy, Chitkara University, Rajpura, Punjab, India; Email: kamya.goyal7@gmail.com

**Anju Goyal & Harish Kumar (Eds.)**
**All rights reserved-© 2022 Bentham Science Publishers**

# INTRODUCTION

In our analytical chemistry, Kjeldahl method is used for the quantitative estimation of nitrogen, which is contained in organic, and as well as inorganic compound ammonia ($NH_3$), or ammonium substances ($NH_4^+$). If the inorganic nitrogen was not modifying for instance nitrate, then it will not be included in this depth [1]. Nitrogen is present in the organic material, and it is varying from the five major elements which are found as proteins. This method was invented by a Danish chemist Johan Kjeldahl in 1883 [2]. He used this method to estimate the protein amount from a given sample for the large varieties of living being. Also, this method was again revised to find the nitrogen amount in a mixture of multiples substances which includes, nitrate, organic compounds, and ammonium salts [3]. The Kjeldahl method for the estimation of nitrogen is used in worldwide as a standard method to calculate the amount of protein from a large variety of materials which includes animal or human food, fertilizer, water waste, and fossil fuels [4].

The sulfuric acid was used in the experimentation of this method for the oxidation of compound central basis used in this procedure is the oxidation of the organic compound which contains strong sulfuric acid [5]. When the organic substance is oxidized the carbon containing compound, it will transformed into carbon dioxide, and the hydrogen into water. The nitrogen which is present in the amine groups was practically available in polypeptide chains of containing bonds. Then it will convert into ammonium ions, and after that ammonium ions will dissolve in the oxidizing solution. In last, it is converted into ammonia gas [6].

## Principles

### Digestion

The food sample is weighed into a required quantity and then added into the *digestion flask.* Then digestion of this process starts by heating it in the occurrence of oxidizing compound known as, which helps digest the food material [7]. Also, anhydrous sodium sulfate is used to accelerate the process of reaction to reach the boiling point, or a catalyst includes copper, selenium, titanium, and mercury are also used to accelerate the rate of reaction.

The whole process of digestion will convert the nitrogen in the food material into ammonia, and other organics (which is mainly present in the form of nitrates, or nitrites) into ammonia, and other organic substances to Carbon dioxide, and water [8]. Ammonia gas is not active in an acid solution because the ammonia is in the form of the ammonium ion ($NH_4^+$) which binds to the sulfate ion ($SO_4^{2-}$) and thus remains in solution:

$$N\ (food) \longrightarrow (NH_4)_2SO_4 \tag{1}$$

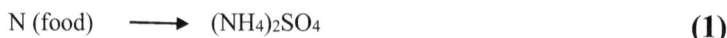

## Neutralization

After the completion of digestion process the digestion flask is connected with the receiving flask with a glass tube [9]. In the process of neutralization the digestion flask made the solution in alkaline form by adding sodium hydroxide (NaOH), which changed the ammonium sulfate into ammonia gas:

$$(NH_4)_2SO_4 + 2\ NaOH \longrightarrow 2NH_3 + 2H_2O + Na_2SO_4 \tag{2}$$

The liberated ammonia gas from digestion flask was expelled out, and entered in to the receiving flask producing the intemperance of boric acid [10]. Due to low value of solution pH the ammonium gas was converted into ammonia, and at the same time it converted the boric acid into borate ions:

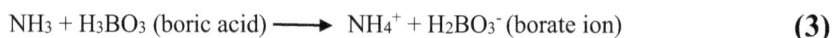

$$NH_3 + H_3BO_3\ (boric\ acid) \longrightarrow NH_4^+ + H_2BO_3^-\ (borate\ ion) \tag{3}$$

## Titration

To estimate the amount of nitrogen the process of titration was started by titrating the nitrogen content by titration process the ammonium, and borate ions were formed with standard sulfuric or hydrochloric acid [11]. A suitable indicator was used to find the end point of the reaction.

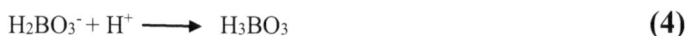

$$H_2BO_3^- + H^+ \longrightarrow H_3BO_3 \tag{4}$$

The hydrogen ions concentration (in moles) was necessary to get the end point which is correspondent to the concentration of nitrogen present in the original food material (see equation no. 3) [12]. To determine the concentration of nitrogen in sample the following equation was used, which contained weight (m in grams) by using a solution xM HCL for the titration process:

$$\%\,N = \frac{x\ moles}{1000\ cm^3} \times \frac{(vs - vb)cm^3}{mg} \times \frac{14\ g}{moles} \times 100$$

Where, $vs$ and $vb$ are the sample and blank volume for titration, and 14g is denoted as the molecular weight of nitrogen (N). Once the nitrogen content has been determined it is converted to a protein content using the appropriate conversion factor: %Protein = F%N [13].

## Procedure

The Kjeldahl method is experimented by three steps, and these steps was applied carefully in a sequence way during the whole procedure (Fig. **1**):

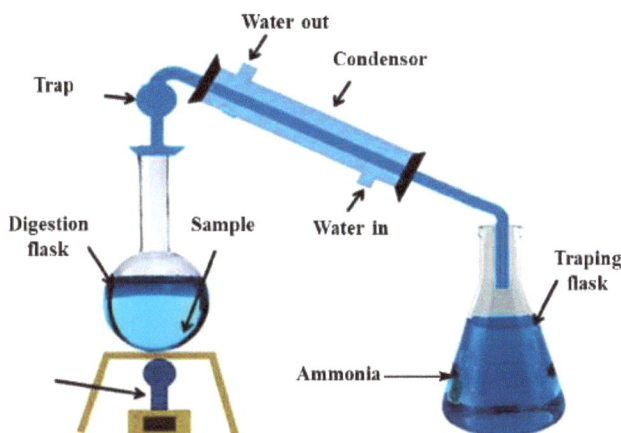

**Fig. (1).** Assembly of Kjeldahl Method.

1. The sample containing food material was added carefully in round volumetric flask which contains sulfuric acid. This system helps the solution to get digest in the occurrence of catalysts, this way helps the material to change the amine nitrogen into ammonium ions [14].

2. In the second step the ammonium ions are next transformed into ammonia gas, by heating and distillation [15]. The ammonia gas is flowing into a trap by mean for flowing the solution in a controlled way, in next the ions get dissolve properly, form once again ammonium ions.

3. In last step the total amount of ammonia has been moved during experimentation is estimated with the help of titration process by standard solution, and the calculations were made [16].

## Step One: Digestion of the Sample

### Digestion is Accomplished by

1. Accurately weighted 1 gm of the sample material which includes protein, and then place the digestion flask on assembly. The digestion flask was filled with concentrated sulfuric acid of 12-15 mL.

2. Now, add 7 gm of catalyst (copper) and potassium sulfate.

3. The solution which brings in digestion flask started boiling with the help of heating source at 370-400 °C temperature. Constantly heat the flask until the fumes are formed, this procedure should be continued for 60-90 min.

4. In last cool the flask by adding 250 ml of water [17].

## Step Two: Distillation

This step was used to separate the ammonia (nitrogen), which is present in digestion flask. This process is done by the following steps [18]:

1. The sodium hydroxide in the concentration of 45% was used to raise the pH of solution. This process led to change the ammonium ions to ammonia in gas form.

2. By distillation the nitrogen gas is separated from digestion flask by ammonia distillation. The ammonia gets converted into volatile gas, until the temperature increases to boiling range. Now, the vapors of distilled water are trapped in a solution containing 15 mL HCl with 70 ml of water.

3. After this process, the flask is removed, and the condenser is rinsed with the help of water, which confirms that the available ammonia in flask was dissolved.

## Step Three: Titration

The Quantities of Acid, and Hence Ammonia is Determined by [19].

1. Add an indicator dye into the trapped solution of HCl/ammonia, and a sharp color was observed which shows that a particular quantity of acid is present in original trapped solution.

2. Now, add a usual solution of NaOH (sodium hydroxide) in the buret tube, and also add small drops of sodium hydroxide solution in a slow way continuously which is mixed the acid solution and dye.

3. Till the end point the dye changes their color to orange, which shows the endpoint. This result indicates that the acid has been neutralized by the base.

4. The volume of the sodium hydroxide solution (neutralized base) was essential to get the endpoint.

5. In last the quantity of ammonia, nitrogen was calculated which was extracted from the original sample.

## Applications

1. The Kjeldahl method is used to estimate the nitrogen in chemical substances, and also this method is used to determine the nitrogen in organic, and inorganic compounds in pharmaceutical industries [20].

2. It is a unique method to determine the content of protein in food, and other substances.

3. It is used to assay soils, fertilizers, waste waters, and other materials.

## Recent Developments

1. In malevolence of all the new developments in the instrumentation, there is an advance system for sample digestion has been implemented to conquer the drawbacks regarding the use of catalyst [21].

2. To manage the efficacy rate in laboratories for the estimation of data an advance procedure to assure the quality department include quality assurance, and quality control have been implemented.

3. The new ideologies have been made with regards digestion process in micro wave for the estimation of ammonium bisulfate were investigated to make sure the future of the Kjeldahl method with the help of combustion nitrogen analyzer technology.

4. There are many alternatives to approach the total amount of nitrogen content which includes alkaline persulfate digestion technique. Due to this procedure the sample of nitrogen will oxidize in the form of nitrate as a sole product, by the use of potassium persulfate under high temperature, and pressure with strong alkaline environment [22].

## CONCLUSION

The Kjeldahl method is extremely versatile, as it can handle a wide range of samples from the food and feed industry (grain, meat, fish, milk, dairy, fruit, vegetables), beverage industry, environmental industry (agriculture, oilseeds, soil, fertilizers, water, wastewater, sludge), and chemical and pharmaceutical industries (paper, textiles, rubber, plastic, polymer). The sample preparation phase is often the key to an effective analysis *via* Kjeldahl method. This approach involves changing the process of oxidizing substances that include carbon dioxide or hydrogen atoms in water. When ammonium ions dissolve in an oxidized solution, they are converted into ammonia gas. Its key concepts are digestion, titration, and neutralization. The three-step principle is utilized to describe the nitrogen estimation process. This approach is highly useful in estimation of nitrogen.

## CONSENT FOR PUBLICATION

Not applicable.

## CONFLICT OF INTEREST

The authors declare no conflict of interest, financial or otherwise.

## ACKNOWLEDGEMENTS

Declared none.

## REFERENCES

[1]     Adams, C.I.; Spaulding, G.H. Determination of organic nitrogen by Kjeldahl method without distillation. *Anal. Chem.,* **1955**, *27*(6), 1003-1004.
[http://dx.doi.org/10.1021/ac60102a040]

[2]     Domini, C.; Vidal, L.; Cravotto, G.; Canals, A. A simultaneous, direct microwave/ultrasound-assisted digestion procedure for the determination of total Kjeldahl nitrogen. *Ultrason. Sonochem.,* **2009**, *16*(4), 564-569.
[http://dx.doi.org/10.1016/j.ultsonch.2008.12.006] [PMID: 19157952]

[3]     Ajaz, A.G. Estimation of ammonium perchlorate in HTPB based composite solid propellants using Kjeldahl method. *J. Hazard. Mater.,* **1995**, *42*(3), 303-306.
[http://dx.doi.org/10.1016/0304-3894(95)00009-J]

[4]     Ali, A.G.; Lovatt, C.J. Evaluating analytical procedures for quantifying ammonium in leaf tissue. *J. Am. Soc. Hortic. Sci.,* **1995**, *120*(5), 871-876.
[http://dx.doi.org/10.21273/JASHS.120.5.871]

[5]     Baethgen, W.E.; Alley, M.M. A manual colorimetric procedure for measuring ammonium nitrogen in soil and plant Kjeldahl digests. *Commun. Soil Sci. Plant Anal.,* **1989**, *20*(9-10), 961-969.
[http://dx.doi.org/10.1080/00103628909368129]

[6]     Lei, C.X. The Problems That Should Pay Attention to in Kjeldahl Methods for Determination of Protein. In: *Cereals Oils Technol*; , **2003**; 1, pp. 62-64.

[7]     Marcali, K.; Rieman, W., III Kjeldahl determination of nitrogen. Elimination of the distillation. *Ind. Eng. Chem. Anal. Ed.,* **1946**, *18*(11), 709-710.
[http://dx.doi.org/10.1021/i560159a016]

[8]     Engler, J.; Kathe, U.; Georgi, E. A Study on the Postreactive Selenium Elimination from Solutions Remaining after Kjeldahl Digestion by Means of Selenium. *Nahrung,* **1986**, *30*(1), 25-30.
[http://dx.doi.org/10.1002/food.19860300108]

[9]     Christensen, L.M.; Fulmer, E.I. A modified Kjeldahl method for the determination of the nitrogen content of yeast. *Plant Physiol.,* **1927**, *2*(4), 455-460.
[http://dx.doi.org/10.1104/pp.2.4.455] [PMID: 16652539]

[10]    Korn, M.; Dos Santos, W.P.; Korn, M.; Ferreira, S.L. Optimisation of focused-microwave assisted digestion procedure for Kjeldahl nitrogen determination in bean samples by factorial design and Doehlert design. *Talanta,* **2005**, *65*(3), 710-715.
[http://dx.doi.org/10.1016/j.talanta.2004.07.047] [PMID: 18969857]

[11]   Krotz, L.; Cicerci, E.; Giazzi, G. Protein determination in cereals and seeds. *Food Quality & Safety.,* **2008**, *15*(4), 37-39.

[12]   Nelson, D.W.; Sommers, L.E. Determination of total nitrogen in plant material 1. *Agron. J.,* **1973**, *65*(1), 109-112.
[http://dx.doi.org/10.2134/agronj1973.00021962006500010033x]

[13]   Michałowski, T.; Asuero, A.G.; Wybraniec, S. The titration in the Kjeldahl method of nitrogen determination: Base or acid as titrant? *J. Chem. Educ.,* **2013**, *90*(2), 191-197.
[http://dx.doi.org/10.1021/ed200863p]

[14]   Möller, J. Protein analysis revisited. *Focus,* **2010**, *34*(2), 22-23.

[15]   Owusu Apenten, R.K. Kjeldahl method, quantitative amino acid analysis and combustion analysis. Food Protein Analysis, **2002**; pp. 1-42.
[http://dx.doi.org/10.1201/9780203910580.ch1]

[16]   Santos, G.A.; Dos Santos, A.P.; Korndoerfer, G.H. System by near infrared (NIR) for analysis of nitrogen foliar. *Biosci. J.,* **2012**, *28*(1), 83-90.

[17]   Rouch, D.A.; Roginski, H.; Britz, M.L.; Roupas, P. Determination of a nitrogen conversion factor for protein content in Cheddar cheese. *Int. Dairy J.,* **2008**, *18*(2), 216-220.
[http://dx.doi.org/10.1016/j.idairyj.2007.07.004]

[18]   Barbano, D.M.; Clark, J.L.; Dunham, C.E.; Flemin, R.J. Kjeldahl method for determination of total nitrogen content of milk: collaborative study. *J Assoc Off Anal Chem,* **1990**, *73*(6), 849-859.
[http://dx.doi.org/10.4319/lom.2009.7.751]

[19]   Amamcharla, J.K.; Metzger, L.E. Evaluation of a rapid protein analyzer for determination of protein in milk and cream. *J. Dairy Sci.,* **2010**, *93*(8), 3846-3857.
[http://dx.doi.org/10.3168/jds.2009-2959] [PMID: 20655454]

[20]   Thiex, N. Evaluation of analytical methods for the determination of moisture, crude protein, crude fat, and crude fiber in distillers dried grains with solubles. *J. AOAC Int.,* **2009**, *92*(1), 61-73.
[http://dx.doi.org/10.1093/jaoac/92.1.61] [PMID: 19382563]

[21]   Valdes, C.; Andres, S.; Giraldez, F.J.; García, R.; Calleja, A. Potential use of visible and near infrared reflectance spectroscopy for the estimation of nitrogen fractions in forages harvested from permanent meadows. *J. Sci. Food Agric.,* **2006**, *86*(2), 308-314.
[http://dx.doi.org/10.1002/jsfa.2309]

[22]   Zhao, D.; Jai, V.; Farkye, N.Y. Determination of True Proteins in Dairy Products: A Comparative Study between Kjeldahl and Sprint Protein Analyzer. *J. Dairy Sci.,* **2010**, *93* Suppl. 1, 334.

**CHAPTER 9**

# Oxygen Flask Combustion Method

**Kamya Goyal[1,2,\*], Navdeep Singh[1], Shammy Jindal[1], Rajwinder Kaur[2], Anju Goyal[2] and Rajendra Awasthi[3]**

[1] *Laureate Institute of Pharmacy, Jawalamukhi, Himachal Pradesh, India*

[2] *Chitkara College of Pharmacy, Chitkara University, Rajpura, Punjab, India*

[3] *Amity Institute of Pharmacy, Amity University Uttar Pradesh, Noida, Uttar Pradesh, India*

**Abstract:** This method was invented by Wolfgang Schoniger in 1955 and it is useful technique utilized for the estimation and identification of some combusting organic compounds of sulfur or halogen in medicine. These compounds include chlorine, bromine, fluorine, sulfur, *etc*. This technique was also named as Schoniger oxygen-flask technique, and it has been modified to tracing, and analyzing the compounds. The compounds which are organic and some other polymers all these are experimented by this technique in microgram amount.

A 500 ml conical flask of heavy wall was used in this method. The test solution was prepared in two different ways for solid as well as liquid samples. For solid samples the required quantity of selected sample in a specific amount at the center part of filter paper were used. For liquid samples appropriate quantity of absorbent cotton along with filter paper of size (50 mm-length/5 mm-width) was used. The procedure of the determination of chlorine, and bromine, Iodine, Fluorine, and Sulfur were discussed. The chlorine and bromine test was experimented with the help of blank solution in a same quantity to make necessary correction. The Iodine test was experimented with the help of blank solution in a same quantity to make necessary correction. Fluorine test was performed with the required prepared solutions under ultraviolet visible spectrophotometry. Lastly, the sulfur test was performed with required quantity of blank solution in a same manner. This method covers the introduction part along with method of sample preparation, and the important applications of oxygen flask method were studied.

**Keywords:** Applications, Halogen, Method of combustion, Oxygen flask combustion method, Platinum gauge, Preparation of test solution, Procedure, Schoniger oxygen-flask technique, Solid sample, Wolfgang Schoniger.

---

\* **Corresponding Author Kamya Goyal:** Laureate Institute of Pharmacy, Jawalamukhi, Himachal Pradesh, India and Chitkara College of Pharmacy, Chitkara University, Rajpura, Punjab, India; Email: kamya.goyal7@gmail.com

**Anju Goyal & Harish Kumar (Eds.)**
**All rights reserved-© 2022 Bentham Science Publishers**

# INTRODUCTION

Oxygen flask combustion (OFC) method was invented by Wolfgang Schoeniger in 1955. The oxygen flask combustion method is a useful technique used for the estimation, and identification of some combusting organic compounds of sulfur or halogens. These compounds include chlorine, bromine, fluorine, sulfur. All these compounds are experimented by this technique under a flask which is filled with oxygen compound. This technique was also named as Schoniger oxygen-flask technique (Fig. **1**), and it has been modified to tracing, and analyzing the compounds. For the experimentation process, an apparatus containing electrical fire was designed, which was proficient to combusting the material amount upto 100 mg. In the previous discussion, we must say that this technique was established to identify the compounds like chlorine, sulfur, phosphorus, and other organic compounds. Also, the use of this technique was more probably to determine and trace the minor amount of compounds for their scat attention. The compounds which are organic, and some other polymers all these are experimented by this technique in microgram amount [1, 2].

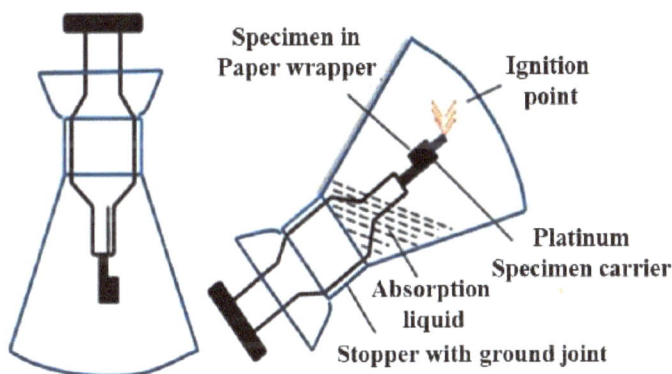

**Fig. (1).** Oxygen Flask Combustion Apparatus.

## Apparatus

The apparatus of oxygen flask combustion method includes conical flask which is a heavy wall of 500 ml volume. In the flask a ground glass stopper was adjusted which is connected with a specimen carrier containing platinum wire made up of heavy gauge. The piece of platinum gauge is a welded piece, and its area is about $1.5 \times 2$ cm (Fig. **1**) [3].

## Preparation of Test Solution and Blank Solution

The samples of test solution were prepared by the following methods below:

### *Preparation of Sample*

- For solid samples: Add the required quantity of selected sample in a specific amount at the center part of filter paper. After that, accurately weight, and carefully wrap the sample without scattering. Now, place the whole packet in a cylinder or basket which is made up of platinum and the fuse strip was leaving outside [4].
- For liquid samples: Spin the appropriate quantity of absorbent cotton along with filter paper of size (50 mm-length/5 mm-width). The last part of the paper was left with the length of the paper in the range of 18-20 mm. Now, place the sample in basket of platinum. Also add the appropriate sample in a glass tube, which is accurately weighed, and a cotton is moistened with specific quantity of sample. Then, the samples start bringing to edge by making contact with cotton [5].

## Method of Combustion

The liquid which is adsorbed properly was placed in a suitable flask, which is filled with oxygen, and the ground part of the stopper was moistened with the help of water. Then the ignition of fuse strip was starts absorbing the specific liquid in to the flask. Straight away the samples were transferred into the flask, and the kept aside the flask by properly air-tightening till the combustion be completed. Intermittently shake the flask continuously till the smoke of white colour was not removed completely. Now, place the samples to stand for 15-30 min, and the vacant solution designating the resulting solution as the test solution. Prepare the blank solution in the same manner, without sample [6].

### Procedure of Determination

#### *Chlorine and Bromine*

Little quantity of water was applied to the upper part of flask, also the point ground stopper was carefully pulled away, and the test solution was transferred into a beaker. The inner side of flask, along with platinum specimen carrier or ground stopper of 2- propanol in 15 mL solution was combined [7].

One drop of bromophenol blue TS dye in the solution was added, also dilute nitric acid was added dropwise until a yellow color was formed. After that 25 mL of 2-propanol and titrate with 0.005 mol/L silver nitrate were added in accordance with the potentiometric titration inside the electrometric titration. This test was experimented with the help of blank solution in a same quantity to make necessary correction.

Each mL of 0.005 mol/L silver nitrate VS = 0.17727 mg of Cl
Each mL of 0.005 mol/L silver nitrate VS = 0.39952 mg of Br

## 2. Iodine

Little quantity of water was applied to the upper part of flask, also the point ground stopper was carefully pulled away [8, 9].

Now, two drops of hydrazine hydrate into the test solution, and then place the ground stopper into flask. Then the solution was decolorized by continuous shaking. The material inside the flask was transferred into a beaker. Then wash the inner side of flask, platinum specimen carrier, and ground stopper with 25 mL of propanol, and the washed material were transferred into another beaker. Now, add one drop of bromophenol blue TS dye in this solution, also add dilute nitric acid dropwise until a yellow color formed. After that add 25 mL of 2-propanol and titrate with 0.005 mol/L silver nitrate in accordance with the potentiometric titration inside the electrometric titration. This test was experimented with the help of blank solution in a same quantity to make necessary correction.

Each mL of 0.005 mol/ L silver nitrate VS = 0.6345 mg of 1

## 3. Fluorine

A small amount of water was applied into the upper part of flask, and the ground stopper was pulled with awareness. Then the test solution was transferred into a volumetric flask containing 50 ml blank solution, and the platinum specimen carrier or ground stopper along with the inner side of flask were washed carefully, and the volume was make upto 50 ml with distilled water. These solutions were used as test solutions or solution for correction.

As 5 ml of solution was pipetted out equivalent to 5 ml standard solution of fluorine, now separately transfer the solution into 50 ml volumetric flask. Add alizarin complex one mixture TS with buffer solution of acetic acid potassium acetate.

The pH of buffer solution were adjusted at 4.3, and then a cerium (III) nitrate TS in the ratio of (1:1:1) were used and the final volume was makeup with water upto 50 ml. This solution was kept to allow stand till to 1 hr. Now, this test was performed with the required prepared solutions under ultraviolet visible spectrophotometry. A 5 ml of blank solution was used with same quantity of water under the UV, and the absorbance was estimated [10].

$A_T$, $A_c$ and $A_s$ are the succeeding solutions of the test solution, the correction, and the standard solution were analyzed at 600 nm.

Amount of fluorine (mg) in the test solution = Amount of fluorine (mg) in ml of standard solution

$$\times \frac{AT-AC}{AS} \times \frac{50}{V}$$

## *Standard Fluorine Solution*

The standard reagent of sodium fluoride was dried under platinum crucible in the temperature range of 500 – 550 °C for 1 hr, and after that cool the solution under desiccator which contains silica gel. 66.5 mg of sample was weighed accurately, and dissolved with water to make the volume upto 500 ml. A 10 ml of solution was pipette out and dilute this solution with required quantity of solution upto 100 ml [11].

## *4. Sulfur*

A little amount of water was applied carefully in the upper part of flask, and ground stopper was carefully pulled away. Now ground stopper, platinum specimen carrier, along with the inner side of flask was washed gently with 15 ml of methanol. Then, add 25 ml of 0.005 mol/L quantity of barium perchlorate, allow this solution stand for 10 min. After that add 0.15 ml of arsenazo III TS with a measuring pipette were added and titrated with 0.005 mol/L sulfuric acid VS. In last, this test was performed with required quantity of blank solution in a same manner [12].

Each mL of 0.005 mol/L barium perchlorate VS = 0.1603 mg of S

## Applications

1. This method was used to estimate or determine the sulfur, and halogen compounds in medicines.

2. Oxygen flask combustion is a suitable method for the estimation of some elements in tablets, capsules, solutions, lozenges, creams, and ointments.

3. Also, this method is suitable for the determination of compounds like bromine, fluorine, iodine *etc*.

4. Oxygen flask method is used for the determination of mercury in coals [13, 14].

## CONCLUSION

Wolfgang Schöniger devised this approach in 1955 and is a valuable tool for estimating and identifying some combusting organic sulphur or halogen compounds in medicine. This approach is used to analyze inorganic compounds as well as certain other polymers in microgram quantities. The advantages of the oxygen flask method are especially evident in sulphur, iodine, and phosphorus measurements. The fact that an effective titrimetric measurement of sulphate was found practically concurrent with the widespread use of the oxygen flask method assisted the sulphur method significantly. Iodine and phosphorus procedures are also far simpler than any of their predecessors. The method's simplicity is perhaps its biggest strength and attempts to complicate it excessively by using electrical ignition, reverting to the period of mechanical failure, must be avoided.

## CONSENT FOR PUBLICATION

Not applicable.

## CONFLICT OF INTEREST

The authors declare no conflict of interest, financial or otherwise.

## ACKNOWLEDGEMENTS

Declared none.

## REFERENCES

[1]   Haslam, J.; Hamilton, J.B.; Squirrell, D.C. The determination of chlorine by the oxygen flask combustion method: a single unit for electrical ignition by remote control and potentiometric titration. *Analyst (Lond.),* **1960**, *85*(1013), 556-560.
      [http://dx.doi.org/10.1039/an9608500556]

[2]   Macdonald, A.M. The oxygen flask method. A review. *Analyst (Lond.),* **1961**, *86*(1018), 3-12.
      [http://dx.doi.org/10.1039/an9618600003]

[3]   Sykes, A. The determination of metals in organic compounds. *Mikrochim. Acta,* **1956**, *44*(7-8), 1155-1168.
      [http://dx.doi.org/10.1007/BF01257449]

[4]   Terentiev, A.P.; Obtemperanskaya, S.I.; Likhosherstova, V.N. Bestimmung von Kupfer in organischen Verbindungen. *Zh. Analit. Khim.,* **1960**, *15*, 748.

[5]   Belcher, R.; West, T.S. Mercurous nitrate as a reductimetric reagent. *Anal. Chim. Acta,* **1951**, *5*, 260-267.

[http://dx.doi.org/10.1016/S0003-2670(00)87545-5]

[6]     Geng, W.; Nakajima, T.; Takanashi, H.; Ohki, A. Utilization of oxygen flask combustion method for the determination of mercury and sulfur in coal. *Fuel,* **2008**, *87*(4-5), 559-564.
        [http://dx.doi.org/10.1016/j.fuel.2007.02.026]

[7]     Haslam, J.; Hamilton, J.B.; Squirrell, D.C. The detection of "additional elements" in plastic materials by the oxygen flask combustion method. *Analyst (Lond.),* **1961**, *86*(1021), 239-248.
        [http://dx.doi.org/10.1039/AN9618600239]

[8]     Lalancette, R.A.; Lukaszewski, D.M.; Steyermark, A. Determination of iodine in organic compounds employing oxygen flask combustion and mercurimetric titration. *Microchem. J.,* **1972**, *17*(6), 665-669.
        [http://dx.doi.org/10.1016/0026-265X(72)90134-8]

[9]     Dobbs, H.E. Oxygen Flask Method for the Assay of Tritium-, Carbon-14-, and Sulfur-35-Labeled Compounds. *Anal. Chem.,* **1963**, *35*(7), 783-786.
        [http://dx.doi.org/10.1021/ac60200a007]

[10]    Manning, A.; Keeling, R.F. Global oceanic and land biotic carbon sinks from the Scripps atmospheric oxygen flask sampling network. Tellus B Chem. *Phys. Meteorol.,* **2006**, *58*(2), 95-116.

[11]    Haslam, J.; Hamilton, J.B.; Squirrell, D.C. Application of the oxygen flask combustion method to the determination of chlorine in polymers, plasticizers and organic compounds. *J. Appl. Chem. (Lond.),* **1960**, *10*(2), 97-100.
        [http://dx.doi.org/10.1002/jctb.5010100208]

[12]    Steyermark, A. Progress in elemental quantitative organic analysis: 1958. *Microchem. J.,* **1959**, *3*(3), 399-414.
        [http://dx.doi.org/10.1016/0026-265X(59)90009-8]

[13]    McGillivray, R.; Woodger, S.C. The application of the oxygen-flask combustion technique to the determination of trace amounts of chlorine and sulphur in organic compounds. *Analyst (Lond.),* **1966**, *91*(1087), 611-620.
        [http://dx.doi.org/10.1039/an9669100611]

[14]    Juvet, R.S.; Chiu, J. Determination of Carbon in Organic Substances by Oxygen-Flask Method. *Anal. Chem.,* **1960**, *32*(1), 130-131.
        [http://dx.doi.org/10.1021/ac60157a039]

CHAPTER 10

# Precipitation Titration

**Kamya Goyal[1,2,\*], Navdeep Singh[1], Shammy Jindal[1], Rajwinder Kaur[2], Anju Goyal[2] and Rajendra Awasthi[3]**

[1] *Laureate Institute of Pharmacy, Jawalamukhi, Himachal Pradesh, India*

[2] *Chitkara College of Pharmacy, Chitkara University, Rajpura, Punjab, India*

[3] *Amity Institute of Pharmacy, Amity University Uttar Pradesh, Noida, Uttar Pradesh, India*

**Abstract:** Precipitation titration is a titration where the precipitation responses occur. In this method of titration, the titrant responds with the analytes which are framed with insoluble material and the process of titration starts with devoured analytes. To decide the convergence of chloride particle in a specific arrangement, we can titrate this arrangement with silver nitrate. In the principle, the contingent on the upsides of ionic item, the arrangements can be characterized into three distinct classes where the precipitations occur. In precipitation titration, generally three methods were studied including Volhard's method which was given by Jacob Volhard in 1874. The scientist planned the technique for assessment of silver particles [AgNO$_3$] in weakening corrosive arrangements by titrating against a standard thiocyanate arrangement within the sight of ferric salt Ferric ammonium sulfate as pointer. The second one is the Mohr's method which was given by Karl Friedrich Mohr in 1856 who presented it for the assurance of halide/chloride with silver nitrate utilizing potassium chromate arrangement as pointer. It is particularly helpful for the assurance of chloride. The third method is Fajan's method which was given by Kazimierz Fajan in 1923 in which the adsorption markers were used. In the measure of substances which respond with nitrate however this can't be controlled by direct titration with silver nitrate arrangement and the greater part of the utilizations of precipitation titrations depend on the utilization of a standard silver nitrate solution. In this chapter, we studied the basic fundamentals of precipitation titration along with their principle, methods, limitations and their suitable application.

**Keywords:** Analyte, Argentometry, Chloride, Complex formation, Corrosive, Fajan's method, Indicators, Limitations, Limits of Precipitation titration, Mohr's method, Principle, Precautions, Preparation of 0.1 M AgNO$_3$, Precipitates, Precipitation titration, Silver nitrate, Solvent, Temperature, Titrant, Volhard's method.

\* **Corresponding Author Kamya Goyal:** Laureate Institute of Pharmacy, Jawalamukhi, Himachal Pradesh, India and Chitkara College of Pharmacy, Chitkara University, Rajpura, Punjab, India; Email: kamya.goyal7@gmail.com

**Anju Goyal & Harish Kumar (Eds.)**
**All rights reserved-© 2022 Bentham Science Publishers**

# INTRODUCTION

An exceptional sort of titrimetric method includes the development of encourages throughout titration. The titrants respond with the analyte framing an insoluble material and the titration proceeds till the absolute last measure of analyte is devoured [1]. The primary drop of titrant in overabundance will respond with a marker bringing about a shading change and reporting the end of the titration. Quantitative precipitation can be utilized for volumetric assurance.

$$AgNO_2 + NaCl \rightarrow AgCl + NaNO_2$$

Precipitation titration is the titration where precipitation responses occur are called precipitation titration [2]. For *e.g.* : titration of $AgNO_3$ with halide particles, for example, Cl or I, Br Solubility item in the accompanying response the result of molar centralization of particles raised to the force equivalent to its stoichiometric coefficient presents in the ionic condition in soaked arrangement at a fixed temperature is called solvency item [3].

## Precipitation Titration Example

To decide the convergence of chloride particle in a specific arrangement, we can titrate this arrangement with silver nitrate arrangement (whose focus is known) [4]. The substance response happens as follows:

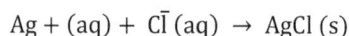

$$Ag + (aq) + \bar{Cl} (aq) \rightarrow AgCl (s)$$

AgCl as a white hasten can be seen settled at the lower part of the cup during titration. The amount of silver particle used to proportionality point is equivalent to the amount of chloride particle which was initially present [5]. To ascertain the quantity of moles of chloride particle or silver particle, we can utilize

$$n = cV \ldots \text{(molarity definition)}$$

To ascertain the volume of the additional arrangement or molar centralization of particle the relating upsides of both of the particles ought to be known.

## PRINCIPLE OF PRECIPITATION

The result of groupings of particles presents in an answer at any fixed temperature is called ionic result of the salt. It is indicated by Q and if the arrangement is soaked arrangement, then the ionic item is called dissolvability result of the salt "if the ionic result of the salt is more prominent than its solvency item esteem, then the precipitation of the salt happens in any case, this is known as the standard of precipitation [6].

as follows:

i. Q= Ksp, the arrangement is simply soaked and no precipitation happens.
ii. If Q>Ksp, the arrangement is supersaturated and precipitation happens
iii. If Q<Ksp, the arrangement is unsaturated and a greater amount of the solute can break down, therefore, no precipitation happens at this condition [7].

## FACTORS INFLUENCING THE SOLUBILITY OF PRECIPITATE

### a) Effect of Temperature

With expanded temperature dissolvability of accelerate increments

### b) Effect of Solvent

solubility of inorganic salt is decreased by expansion of natural solvent like ethanol, methanol, propanol.

- But in presence of just water, hydration of particles of salt increments because of the great dipole snapshot of water atom.
- This hydration produces energy called hydration energy which is adequate to defeat the appealing power between particles of strong cross section.
- The particles in precious stones don't have so huge a fascination for natural solvents, and thus the solubility are generally not exactly in water [8, 9].

### c) Effect of Corrosive

The solvency of the salt of feeble corrosive is influenced by the expansion of corrosive. Hydrogen particle of added corrosive joins with the anions of the salt and structures feeble corrosive subsequently expanding the solvency of the salt [10].

### d) Formation of Complex Particles

The expansion in solvency of an encourage after adding abundance hastening specialist is much of the time because of the development of complex particle [11].

# METHODS FOR PRECIPITATION TITRATION

- Mohr's Method
- Volhard's Method
- Fajan's Method

## Mohr's Method

In 1856, Karl Friedrich Mohr presented it for the assurance of halide – chloride with silver nitrate utilizing potassium chromate arrangement as pointer. It is particularly helpful for the assurance of chloride.

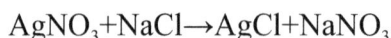

$$AgNO_3+NaCl\rightarrow AgCl+NaNO_3$$

Hastened silver chromate, through sparingly solvent in water is more dissolvable than silver chloride and the red tone because of silver chromate doesn't show up until all the chloride has been accelerate as silver chloride [12].

The affectability of marker relies upon focus, temperature $H^+$ particle fixation, convergence of electrolyte and way of noticing the red shading.

This strategy uses chromate as a marker. Chromate frames a precipitate with $Ag^+$ however this encourage has a more noteworthy solvency than that of AgCl, for instance. Subsequently, AgCl is shaped first and after all $Cl^-$ is devoured, the main drop of $Ag^+$ in abundance will respond with the chromate marker giving a ruddy hasten [13].

$$2Ag^++Cr_4^{2-}\rightarrow Ag_2CrO_4$$

## *Precautions*

In this technique, unbiased medium ought to be utilized since, in antacid arrangements, silver will respond with the hydroxide particles shaping AgOH. In acidic arrangements, chromate will be changed over to dichromate. In this way, the pH of arrangement ought to be kept at around 7. There is in every case some blunder in this strategy in light of the fact that a weaken chromate arrangement is utilized because of the extraordinary shade of the marker. This will require extra measure of $Ag^+$ for the $Ag_2CrO_4$ to shape [14].

## *Limitations*

- Allowable PH range is 6.5 to 10.
- Below pH 6.5 there is expanded in dissolvability of silver chromate.
- Above pH 10 the endpoint comes past the point of no return and Silver

hydroxide is additionally hastened.

- If the arrangement is basic make it acidic with nitric corrosive at that point kill it by adding sodium bi carbonate or borax.
- If considerable measure of Ammonium salts is available, the pH should not surpass 7.2.
- In switch titration iodides and bromides can't be titrated [15].

## *Preparation of 0.1 M Silver Nitrate*

Gauge 17 g of silver nitrate broke up it in 1000 ml of refined water. Weigh precisely 0.1 g of sodium chloride disintegrate in 5 ml of water, 5 ml of acidic corrosive, 50 ml of methanol, 0.15 ml of eosin stir ideally with attractive stirrer and titrate with silver nitrate. End point appearance of pink tone is (Rose milk colour) [16].

## Volhard Method: Jacob Volhard (1834-1910)

In 1874, Jacob Volhard planned the technique for assessment of silver particles [$AgNO_3$] in weakens corrosive arrangements by titrating against a standard thiocyanate arrangement within the sight of ferric salt Ferric ammonium sulfate as pointer. It has been stretched out to assess chloride, bromide and other a few examinations [17].

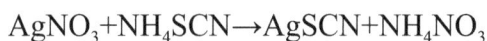

$$AgNO_3 + NH_4SCN \rightarrow AgSCN + NH_4NO_3$$

Ammonium or potassium thiocyanate arrangement is utilized related to 0.1 M $AgNO_3$ in the test of substances which respond with nitrate, yet which can't be dictated by direct titrations with silver nitrate arrangement [18].

- In this technique to the halide arrangement, a known abundance of silver nitrate is titrated with 0.1M smelling salts or potassium thiocyanate arrangement is called Volhard's strategy.
- In this technique, the hasten of silver chloride is separated off and the filtrate is titrated with standard thiocyanate arrangement utilizing ferric ammonium sulfate arrangement as marker. At the endpoint a lasting red tone is delivered because of the development of ferric thiocyanate [19].
- This is an aberrant technique for chloride assurance where an abundance measure of standard $Ag^+$ is added to the chloride arrangement containing $Fe^{3+}$ as a pointer. The overabundance $Ag^+$ is then titrated with standard $SCN^-$ arrangement until a red tone is acquired which results from the response:

$$Fe^{3+} + SCN^- \rightarrow FeSCN^{2+}$$

## Fajan's Strategy: Karl Kazimierz Fajan (1887-1975)

- In 1923-24 Fajan presented the strategy.
- Adsorption marker is utilized.
- The activity of these markers depends on the straightforward truth that the endpoint the pointers get adsorbed by the encourage [AgCl] and during the cycle of adsorption, an adjustment of shade of the pointer will happen which may bring about a substance of various shading [20].

Fluoresce in and its subordinates are adsorbed to the outside of colloidal AgCl. After all chloride is utilized, the primary drop of $Ag^+$ will respond with fluorescein ($FI^-$) framing a ruddy tone.

$$Ag^+ + FI^- \rightarrow AgF$$

## LIMITS OF PRECIPITATION TITRATION

- A few numbers of particles, for example, halide ions ($Cl^-$, $Br^-$, $I^-$) can be titrated by precipitation technique.
- Co-precipitation might happen.
- It is extremely hard to distinguish the end point [21].

## HOW TO CONQUER THE ISSUES OF PRECIPITATION TITRATION

- In the measure of substances which respond with nitrate however this can't be controlled by direct titration with silver nitrate arrangement.
- Excess standard silver nitrate arrangement is added with concentrated nitric corrosive and the abundance silver nitrate titrated with 0.1N ammonium thiocyanate arrangement. (This is called Volhard Method).
- In the instance of chloride, it is typically sifting off the AgCl or to coagulate the hasten through nitrobenzene, which is non-harmful, on the grounds that AgCl responds gradually with ammonium thiocyanate [22].

**Sorts of Medication are Examined by this Technique:** Carbromal, KCl Infusion, NaCl Infusion, Thiamine Hydrochloride.

**Marker of Precipitation Titration:** Potassium Chromate ($K_2CrO_4$) Silver Chromate ($Ag_2CrO_4$)

## APPLICATIONS

The greater part of the utilizations of precipitation titrations depend on the utilization of a standard silver nitrate solution. Table **1** records some ordinary utilizations of argentometry [23].

**Table 1. Typical Argentometric Precipitation Method.**

| Substance Determined | End Point | Remarks |
|---|---|---|
| $AsO_4^{3-}$, $Br^-$, $I^-$, $CNO^-$, $SCN^-$, | Volhard | Removal of silver salt not required |
| $CO_3^{2-}$, $CrO_4^{2-}$, $CN^-$, $Cl^-$, $C_2O_4^{2-}$, $PO_4^{3-}$, $S^{2-}$, $NCN^{2-}$ | | Removal of silver salt required before back–titration of excess $Ag^+$ |
| $BH_4^-$ | | Titration of excess $Ag^+$ following $BH_4^- + 8Ag^+ + 8OH^-\ 8Ag(s) + H_2BO_3^- + 5H_2O$ |
| Epoxide | | Titration of excess $Cl^-$ following hydrohalogenation |
| $K^+$ | | Precipitation of $K^+$ with known excess of $B(C_6H_5)_4^-$ addition of excess $Ag^+$ giving $AgB(C_6H_5)_4(s)$ and back– titration of the excess |
| $Br^-$, $Cl^-$ | Mohr | - |

- In large numbers of these strategies, the analyte is hastened with a deliberate overabundance of silver nitrate, and this progression is then trailed by Volhard titration with standard potassium thiocyanate [24].
- Both silver nitrate and potassium thiocyanate are possible in essential standard quality [25].
- The last is, in any case, fairly hygroscopic and thiocyanate arrangements are commonly normalized against silver nitrate [26, 27].
- Both silver nitrate and potassium thiocyanate arrangements are steady inconclusively [28, 29].
- Silver nitrate is a remarkably useful reagent for both titrimetric and gravimetric methods of analysis. Its greatest drawback is its cost, which fluctuates widely with the world price of silver [30, 31].

## CONCLUSION

Precipitation titration is a type of titration in which precipitation responses are measured. The titrant replies with analytes that are framed with insoluble material in this type of titration, and the titration process begins with absorbed analytes. In the study of precipitation titration, three approaches were studied: Volhard's method, Karl Friedrich Mohr method, and Fajan method.

## CONSENT FOR PUBLICATION

Not applicable.

## CONFLICT OF INTEREST

The authors declare no conflict of interest, financial or otherwise.

## ACKNOWLEDGEMENTS

Declared none.

## REFERENCES

[1]     Syunyaev, R.; Balabin, R.; Akhatov, I.; Safieva, J. Adsorption of petroleum asphaltenes onto reservoir rock sands studied by near-infrared (NIR) spectroscopy. *Energy Fuels,* **2009**, *23*(3), 1230-1236.
        [http://dx.doi.org/10.1021/ef8006068]

[2]     Shahebrahimi, Y.; Zonnouri, A. A new combinatorial thermodynamics model for asphaltene precipitation. *J. Petrol. Sci. Eng.,* **2013**, *109*, 63-69.
        [http://dx.doi.org/10.1016/j.petrol.2013.07.013]

[3]     Salahshoor, K.; Zakeri, S.; Mahdavi, S.; Kharrat, R.; Khalifeh, M. Asphaltene deposition prediction using adaptive neurofuzzy models based on laboratory measurements. *Fluid Phase Equilib.,* **2013**, *337*, 89-99.
        [http://dx.doi.org/10.1016/j.fluid.2012.09.031]

[4]     Nakhli, H.; Alizadeh, A.; Moqadam, M.S.; Afshari, S.; Kharrat, R.; Ghazanfari, M. Monitoring of asphaltene precipitation: experimental and modeling study. *J. Petrol. Sci. Eng.,* **2011**, *78*(2), 384-395.
        [http://dx.doi.org/10.1016/j.petrol.2011.07.002]

[5]     Kord, S.; Ayatollahi, S. Asphaltene precipitation in live crude oil during natural depletion: experimental investigation and modeling. *Fluid Phase Equilib.,* **2012**, *336*, 63-70.
        [http://dx.doi.org/10.1016/j.fluid.2012.05.028]

[6]     Moradi, S.; Dabir, B.; Rashtchian, D.; Mahmoudi, B. Effect of miscible nitrogen injection on instability, particle size distribution, and fractal structure of asphaltene aggregates. *J. Dispers. Sci. Technol.,* **2012**, *33*(5), 763-770.
        [http://dx.doi.org/10.1080/01932691.2011.567878]

[7]     Bagheripour, P. Committee neural network model for rock permeability prediction. *J. Appl. Geophys.,* **2014**, *104*, 142-148.
        [http://dx.doi.org/10.1016/j.jappgeo.2014.03.001]

[8]     Ansari, H.R. Use seismic colored inversion and power law committee machine based on imperial competitive algorithm for improving porosity prediction in a heterogeneous reservoir. *J. Appl. Geophys.,* **2014**, *108*, 61-68.
        [http://dx.doi.org/10.1016/j.jappgeo.2014.06.016]

[9]     Naseri, A.; Khishvand, M.; Sheikhloo, A. A Correlations approach for prediction of PVT properties of reservoir oils. *Petrol. Sci. Technol.,* **2014**, *32*(17), 2123-2136.
        [http://dx.doi.org/10.1080/10916466.2010.551815]

[10]    Gholami, A.; Asoodeh, M.; Bagheripour, P. How committee machine with SVR and ACE estimates bubble point pressure of crudes. *Fluid Phase Equilib.,* **2014**, *382*, 139-149.
        [http://dx.doi.org/10.1016/j.fluid.2014.08.033]

[11]    Asoodeh, M.; Gholami, A.; Bagheripour, P. Oil-$CO_2$ MMP determination in competition of neural network, support vector regression, and committee machine. *J. Dispers. Sci. Technol.,* **2014**, *35*(4),

564-571.
[http://dx.doi.org/10.1080/01932691.2013.803255]

[12]   Hemmati-Sarapardeh, A.; Khishvand, M.; Naseri, A.; Mohammadi, A.H. Toward reservoir oil viscosity correlation. *Chem. Eng. Sci.,* **2013**, *90*, 53-68.
[http://dx.doi.org/10.1016/j.ces.2012.12.009]

[13]   Bagheripour, P.; Asoodeh, M. Genetic implanted fuzzy model for water saturation determination. *J. Appl. Geophys.,* **2014**, *103*, 232-236.
[http://dx.doi.org/10.1016/j.jappgeo.2014.02.002]

[14]   Khishvand, M.; Khamehchi, E. Nonlinear risk optimization approach to gas lift allocation optimization. *Ind. Eng. Chem. Res.,* **2012**, *51*(6), 2637-2643.
[http://dx.doi.org/10.1021/ie201336a]

[15]   Hsieh, Y.L.; Ilevbare, G.A.; Van Eerdenbrugh, B.; Box, K.J.; Sanchez-Felix, M.V.; Taylor, L.S. pH-Induced precipitation behavior of weakly basic compounds: determination of extent and duration of supersaturation using potentiometric titration and correlation to solid state properties. *Pharm. Res.,* **2012**, *29*(10), 2738-2753.
[http://dx.doi.org/10.1007/s11095-012-0759-8] [PMID: 22580905]

[16]   Vermeulen, A.C.; Geus, J.W.; Stol, R.J.; De Bruyn, P.L. Hydrolysis-precipitation studies of aluminum (III) solutions. I. Titration of acidified aluminum nitrate solutions. *J. Colloid Interface Sci.,* **1975**, *51*(3), 449-458.
[http://dx.doi.org/10.1016/0021-9797(75)90142-3]

[17]   Korkmaz, D. Precipitation titration: "determination of chloride by the Mohr method. *Methods,* **2001**, *2*(4), 1-6.

[18]   Behrens, M.; Brennecke, D.; Girgsdies, F.; Kißner, S.; Trunschke, A.; Nasrudin, N.; Zakaria, S.; Idris, N.F.; Abd Hamid, S.B.; Kniep, B.; Fischer, R. Understanding the complexity of a catalyst synthesis: Co-precipitation of mixed Cu, Zn, Al hydroxycarbonate precursors for Cu/ZnO/Al$_2$O$_3$ catalysts investigated by titration experiments. *Appl. Catal. A Gen.,* **2011**, *392*(1-2), 93-102.
[http://dx.doi.org/10.1016/j.apcata.2010.10.031]

[19]   Liu, D.; Zhang, C.; Zhou, X.J.; Zhou, F.H.; Wu, L.L. Fast determination of Ba, Ti contents in BaTiO$_3$ electronic ceramic nanopowders by means of precipitation-complexometric titration. *Xinan Minzu Daxue Xuebao. Ziran Kexue Ban,* **2007**, 5. [J].

[20]   Merroun, M.L.; Nedelkova, M.; Ojeda, J.J.; Reitz, T.; Fernández, M.L.; Arias, J.M.; Romero-González, M.; Selenska-Pobell, S. Bio-precipitation of uranium by two bacterial isolates recovered from extreme environments as estimated by potentiometric titration, TEM and X-ray absorption spectroscopic analyses. *J. Hazard. Mater.,* **2011**, *197*, 1-10.
[http://dx.doi.org/10.1016/j.jhazmat.2011.09.049] [PMID: 22019055]

[21]   Eggermont, S.G.; Prato, R.; Dominguez-Benetton, X.; Fransaer, J. Metal removal from aqueous solutions: insights from modeling precipitation titration curves. *J. Environ. Chem. Eng.,* **2020**, *8*(1)103596
[http://dx.doi.org/10.1016/j.jece.2019.103596]

[22]   Moreira, L.A.; Boström, M.; Ninham, B.W.; Biscaia, E.C.; Tavares, F.W. Hofmeister effects: Why protein charge, pH titration and protein precipitation depend on the choice of background salt solution. *Colloids Surf. A Physicochem. Eng. Asp.,* **2006**, *282*, 457-463.
[http://dx.doi.org/10.1016/j.colsurfa.2005.11.021]

[23]   Egorov, V.V.; Repin, V.A. Precipitation potentiometric titration of physiologically active amines using ion-selective electrodes. *J. Anal. Chem.,* **1993**, *48*(12), 1387-1392.

[24]   Naseri, S.; Tatar, A.; Bahadori, M.; Rozyn, J.; Kashiwao, T.; Bahadori, A. Application of radial basis function neural network for prediction of titration-based asphaltene precipitation. *Petrol. Sci. Technol.,* **2015**, *33*(23-24), 1875-1882.
[http://dx.doi.org/10.1080/10916466.2015.1108984]

[25]   Hemmati-Sarapardeh, A.; Dabir, B.; Ahmadi, M.; Mohammadi, A.H.; Husein, M.M. Modelling asphaltene precipitation titration data: A committee of machines and a group method of data handling. *Can. J. Chem. Eng.,* **2019**, *97*(2), 431-441.
[http://dx.doi.org/10.1002/cjce.23254]

[26]   Archer, M.; Brits, M.; Prevoo-Franzsen, D.; Quinn, L. High concentration aqueous sodium fluoride certified reference materials for forensic use certified by complexometric titration. *Anal. Bioanal. Chem.,* **2015**, *407*(11), 3205-3209.
[http://dx.doi.org/10.1007/s00216-014-8229-2] [PMID: 25326884]

[27]   Fernando, Q.; Butcher, J. Calculation of titration error in precipitation titrations: A graphical method. *J. Chem. Educ.,* **1967**, *44*(3), 166.
[http://dx.doi.org/10.1021/ed044p166]

[28]   McCallium, C.; Midgley, D. Improved linear titration plots for potentiometric precipitation and strong acid-strong base titrations. *Anal. Chim. Acta,* **1973**, *65*(1), 155-162.
[http://dx.doi.org/10.1016/S0003-2670(01)80173-2]

[29]   Fiske, C.H.; Logan, M.A. The determination of calcium by alkalimetric titration. 2. The precipitation of calcium in the presence of magnesium, phosphate, and sulfate, with applications to the analysis of urine. *J. Biol. Chem.,* **1931**, *93*(1), 211-226.
[http://dx.doi.org/10.1016/S0021-9258(18)76505-8]

[30]   Dousma, J.; De Bruyn, P.L. Hydrolysis-precipitation studies of iron solutions. I. Model for hydrolysis and precipitation from Fe (III) nitrate solutions. *J. Colloid Interface Sci.,* **1976**, *56*(3), 527-539.
[http://dx.doi.org/10.1016/0021-9797(76)90119-3]

[31]   Kunduz, N.; Secken, N. Development and application of 7E learning model-based computer-assisted teaching materials on precipitation titrations. *J. Balt. Sci. Educ.,* **2013**, *12*(6), 784-792.
[http://dx.doi.org/10.33225/jbse/13.12.784]

# SUBJECT INDEX

**Anju Goyal & Harish Kumar (Eds.)**
**All rights reserved-© 2022 Bentham Science Publishers**

# W

www.ingramcontent.com/pod-product-compliance
Lightning Source LLC
Chambersburg PA
CBHW041714210326
41598CB00007B/642

* 9 7 8 9 8 1 5 0 5 0 2 5 7 *